T0137247

Ajay Kumar Srivastava

Si Detectors and Characterization for HEP and Photon Science Experiment

How to Design Detectors by TCAD Simulation

 Springer

Ajay Kumar Srivastava
Department of Physics
Chandigarh University
Gharuan, Mohali, Punjab, India

ISBN 978-3-030-19533-5   ISBN 978-3-030-19531-1   (eBook)
https://doi.org/10.1007/978-3-030-19531-1

This Springer imprint is published by the registered company Springer Nature Switzerland AG.
The registered company address is: Gewerbestrasse 11, 6330 Cham, Switzerland

*Inner strength becomes so strong
and pervasive
that nothing can shake it.*
—Sri Sri Ravi Shankar Jee

Dedicated to
My Wife,
Anupma Srivastava

Without whom, it would never
have been possible

Preface

Si detector technology is well established in the High-Energy Physics (HEP) experiments for the track reconstruction of ionizing radiations and photon science experiments for X-ray imaging. The innovation in detector technology is increasing day by day, and for that, various upgrade of HEPs and photon science experiments and high-precision detectors are required for long-term operation of the experiments.

The various types of application of Si detectors and basic device physics information can be obtained from the book of Gerhard Lutz on semiconductor radiation detectors. This book is dedicated to the status and development of detectors for the HL-LHC experiment and fourth-generation photon science experiment (European XFEL). The physics and technology that are required to develop radiation hard Si detectors for X-ray-induced surface damage and bulk damage by hadronic irradiations are discussed in detail.

The present book contains 11 chapters in such a way that a researcher and master's level students can understand the physics and technology of Si detectors. Its concepts can be applied to optimize the design using TCAD device simulation for better electrical performance of the detectors which are used in the harsh radiation environment of the colliders (Hl-LHC) and at XFEL like photon science experiments.

For the very first time, we are presenting the first edition of the book that has a kind of aforesaid chapters on the radiation hard (surface and bulk) Si detectors that are required to learn for the development of technology and make hands on the detectors in the TCAD and measurement laboratory.

This book can be purchased by any researcher and Master of Science (Physics) and B.Tech. (Nuclear Science and Technology) students.

Mohali, India   Ajay Kumar Srivastava
March 2019

Acknowledgements

First and foremost, I would like to thank my supervisor, Prof. (Dr.) R. K. Shivpuri. He provided all the financial, emotional and moral support. He is the best advisor and teacher I could have wished for his vast knowledge, deep insights, tremendous experience and easy grasp of physics at its most fundamental level helped me in understanding the basics of particle physics. He completely changed my way of thinking, quest for knowledge and attitude towards life. It was both a privilege and an honour to work with him. He provided me an opportunity to work as a research assistant and scientist in an international project "Search for the new particles at Large Hadron Collide at CERN, Geneva", sponsored by the Department of Science and Technology (DST), Government of India.

I am also obliged to the head of the Department of Physics, late Prof. M. P. Srivastava, for providing me all the necessary facilities in the Department at that time.

I would like to thank the Department of Science and Technology (DST), Government of India, and ISPN 2003 symposium organizers for providing me the travel and other necessary assistance to attend the 1st International Symposium on Point Defects and Non-stoichiometry (ISPN 2003) held on 20–22 March 2003 at Sendai, Japan and the University Grants Commission (UGC), India, for providing me the travel and other assistance to attend the IEEE Nuclear Science Symposium (NSS) held on 16–22 October 2004 at Rome, Italy.

I would like to thank Prof. Brajesh C. Choudhary for his guidance. In the lab, I was fortunate to get some very knowledgeable and friendly people who constantly helped me in my research work. First and foremost, I remember Kirti Ranjan without whose help and encouragement, I would never have reached here. I worked with him on silicon device simulation and hardware work. He accompanied me to several research institutes, like Bhabha Atomic Research Centre (BARC), Mumbai, India; Tata Institute of Fundamental Research (TIFR), Mumbai, India; University of Panjab, Chandigarh, India; and Bharat Electronics Limited (BEL), Bangalore, India. Second, I would like to remember Ashutosh Bhardwaj who worked with me on silicon strip detectors. He helped me in writing all my papers. I have gained a lot by interacting with them and learned from them Keithley Instruments for the I-V and C-V measurements of silicon strip detectors. Thanks to Namrata and other members of the lab, namely, Sudeep Chatterji, Manoj Jha, Sushil Singh Chauhan, Pooja Gupta, Ashish Kumar, Nayeem, Ashutosh Srivastava and Shilpee Arora, for their useful discussions. I wish to thank Mr. P. C. Gupta and Mr. Sharma for their support and general assistance. I also appreciate the continuous assistance that I received from Mr. Rajendra Mishra, Mr. Mohammad Yunus and Mr. Dinesh.

I would also like to express my sincere thanks to our collaborators from BARC, Dr. S. K. Kataria, Mr. M. D. Ghodgaonkar, Mr. V. B. Chandratre, Dr. Anita Topkar, Mr. M. Y. Dixit and Mr. Vijay Mishra, for their assistance, cooperation and valuable annotations at various India-CMS meetings. I also extend my gratitude to our other collaborators from TIFR, Prof. S. N. Ganguli, Prof. Atul Gurtu and Prof. Sunanda Banerjee, and from Panjab University, Prof. J. M. Kohli, Prof. Suman Beri, Dr. Manjeet Kaur and Dr. J. B. Singh, for their encouragement in accomplishing this work. I extend my appreciation to Dr. O. P. Wadhwan and Dr. G. S. Virdi, CEERI, Pilani, and Mr. Subhash Chandran, Mr. Prabhakar Rao and Mr. Shanker Narayan, BEL, Bangalore, for providing me detailed information about the detector fabrication. I would especially like to thank Praveen and Nikhil with whom I worked at BEL on static measurements of silicon sensors.

I am glad to write a lot of thanks to Prof. Jorma Tuominiemi (CMS programme director), Helsinki Institute of Physics (HIP), University of Helsinki, Finland, for providing me a 3-month invitation to work in the CMS programme activities at HIP under "Indo-Finland Cultural Exchange Programme Scholarship Grant". This work has been sponsored by the Centre for International Mobility (CIMO), Finland, and the UGC, India, for 3 months (1 September 2005 to 30 November 2005). I am highly obliged to Dr. Veikko Karimaki, Dr. Eija Tuominen, Dr. Jaakko Harkonen, Dr. Tomas Linden, Lauri Wendland and Dr. Sourov Roy, Helsinki Institute of Physics (HIP), University of Helsinki, Finland, for their useful discussions and suggestions during my stay there.

In writing this book, I have profited from many discussions with our colleagues, in particular from my co-scientist at the Institute of Experimental Physics, University of Hamburg, Germany. I want to mention in particular Eckhart Fretwurst, Robert Klanner, J. Zhang, D. Eckstein, G. Steinbrück, P. Schpler and M. Moll.

Moreover, I would like to express my thanks to my parents Late Durga Prasad Srivastava and Madhur Srivastava for their efforts and sacrifices that they have done for my higher education making this thesis a reality. I am more thankful to my brother, Sanjeev; my sister, Smita; and my father-in-law who provided me a constant support in every manner after the death of my mother. Specially, I would like to thank my lovely wife, Anupma; my daughters, Anshika and Anuska; and my son, Anmol. I would not have been able to do this without their support. They provided me courage and power to complete my book.

It might be possible that I have forgotten the names of few who made this possible, so a big thanks to them too. Some of them are not even alive to see this day. I would like to express my gratitude to all of them.

Ajay Kumar Srivastava

Contents

Chapter 1
Development of Si Detectors for the CMS LHC Experiments

Ever since the dawn of civilization, man's greatest quest has been to understand the fundamental constituents of matter and the laws governing their interactions. Particle physics is the study of the basic constituents of matter and the fundamental forces acting among them.

Physicists are continuously trying to provide the satisfactory answers to open questions in particle physics using High-Energy Physics (HEP) experiments, which are carried out at laboratories scattered all round the world, where sub-atomic particles collide at very high energies creating the particle collisions which are subsequently studied and analyzed. Most of these experiments involve large international collaborations and are performed at international laboratories such as CERN in Geneva, Switzerland; DESY in Hamburg, Germany and Fermilab (FNAL) in Chicago, USA. There is no possible substitute for these experiments, and experimental techniques require NEW IDEAS in both hardware and software and continuous innovation in physics analysis [1].

The large Hadron Collider (LHC) under construction at CERN, Geneva, will be the largest collider in the world, which will offer a future insight in understanding the basic mechanism of nature. The LHC will produce high-energy collisions of proton-proton (p-p) beams at the centre of mass energy ($\sqrt{s}$) of 14 TeV and peak luminosity (L_{peak}) up to ~1×10^{34} cm^{-2} s^{-1} [2–4].

LHC program, due to begin operation in 2008, will address the fundamental questions in particle physics, namely the search of Standard Model (SM) Higgs boson up to the whole possible mass range from 100 GeV–1 TeV, search for Higgs boson in the minimal supersymmetric (MSSM) model, origin of the spontaneous symmetry breaking (SSB) mechanism in the electroweak (EW) sector of the SM, testing grand unification models, search for supersymmetry (SUSY), new gauge bosons, etc.

© Springer Nature Switzerland AG 2019
A. K. Srivastava, *Si Detectors and Characterization for HEP and Photon Science Experiment*, https://doi.org/10.1007/978-3-030-19531-1_1

To exploit the physics capabilities of the LHC, four experiments are planned and each experiment has its own individual detector, reflecting what each experiment is designed to look for: ATLAS-A Toridal LHC Apparatus, CMS-Compact Muon Solenoid, ALICE-A Large Ion Collider Experiment, and LHC-b–Large Hadron Collider beauty [3]. The primary goal for both ATLAS and CMS detector is to find the Higgs boson, ALICE is to exploit the unique physics potential of nucleus-nucleus interactions at LHC, and LHC-b for precise measurements of CP violation and rare decays.

The use of the CMS detector at LHC will allow to accomplish a breakthrough in the investigation of fundamental laws of nature applicable at extremely small distances, which in turn will allow to open a new skyline in the matter structure. The CMS detector is a general-purpose detector designed for the search of the SM Higgs boson in the whole mass range upto 1 TeV and to exploit the physics of p-p collisions over the full range of luminosities expected at the LHC. It will be sensitive to Higgs boson in MSSM model, and well adapted for SUSY particles searches etc. [5].

1.1 CERN

CERN is the European Laboratory for Particle Physics situated in Geneva, Switzerland. It came into existence in 1954 as one of Europe's first joint ventures in particle physics. Figure 1.1 shows the top view of CERN. The Laboratory provides state-of-

Fig. 1.1 Top view of CERN

the-art scientific facilities for researchers. These facilities include accelerators that accelerate the sub-atomic particles almost upto the speed of light and detectors that can detect the particles.

Accelerators
CERN's accelerator complex is built around three principal inter-dependent accelerators. The oldest, the Proton Synchrotron (PS), was built in the 1950s and was briefly the world's highest energy accelerator. The Super Proton Synchrotron (SPS) was built in the 1970s. The Large Electron-Positron collider (LEP) was the Laboratory's last flagship, which came on stream in 1989. LEP was an enormous machine, built in a circular underground tunnel, which is 27 km in circumference. CERN is currently preparing to install a new accelerator inside the same LEP tunnel called the LHC which will start-up in 2008 and provide the world's physicists a new tool to probe deeper than ever into the heart of matter.

1.2   The Large Hadron Collider (LHC) Machine

To find the answer to the most common question in High-Energy Physics (HEP), i.e., "what is the origin of mass?" previous and ongoing experiments such as LEP collider (e^+-e^-) at CERN and the Tevatron collider (p-$\bar{p}$) at Fermi National Accelerator Laboratory (FNAL) collide sub-atomic particles at centre of mass energies up to 180 GeV and 1.8 TeV respectively. So far, HEP experiments support the theory of fundamental particles and forces known as Standard Model (SM). However, the SM is unattractive as it requires many empirical "input parameters" such as masses of the fundamental particles. The SM cannot also explain the predicted masses of the various available fundamental particles. One of the attractive theory which can be accommodated in SM, known as the "Higgs Mechanism" [5, 6], suggests that particles can acquire mass on interacting with "Higgs field". The strength with which a particle interacts with the Higgs field determines the magnitude of the mass of the particle. The carrier of the SM Higgs field is an electro-weak particle known as "Higgs boson". Several attempts have been made to find the SM Higgs boson but so far it has not been found. LEP experiment at CERN and Tevatron collider at FNAL has indicated that the lowest mass of the Higgs (M_H) is around 114 GeV [7]. The upper most bound on Higgs mass is ~1 TeV as given by mathematical "unitarity".

LHC will replace LEP, in keeping the CERN's cost-effective strategy of building on previous investments. It is designed to use the 27 km LEP tunnel. Figure 1.2 shows an overall view of the LHC experiment at CERN, Geneva.

At LHC machine two multipurpose detectors CMS and ATLAS would carry out studies of p-p interactions. The wide range of physics possibilities will enable LHC to retain its unique place on the frontier of physics research in this century.

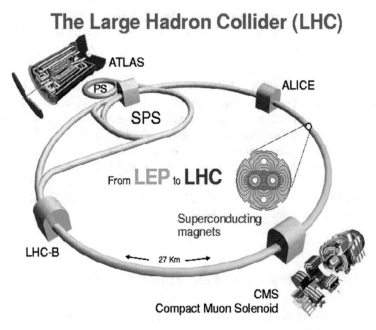

Fig. 1.2 Overall view of LHC experiment at CERN Geneva

1.2.1 Collisions at LHC

LHC will accelerate two proton beams in opposite directions at velocity close to the velocity of light and collide bunches of protons at $\sqrt{s} = 14$ TeV at a frequency of 40 MHz. Figure 1.3 shows the collision schedule at LHC accelerator. The bunches of protons will be focused at the centre of the two multipurpose detectors CMS and ATLAS.

This will occur for the first couple of years at low luminosity (L) $\sim 10^{33}$ cm^{-2} s^{-1}, and will provide on an average $\sim$2–3-proton interactions for every bunch crossing in 25 ns. A high luminosity period will then follow (L $\sim 10^{34}$ cm^{-2} s^{-1}) resulting in an increase of average value of p-p interactions to $\sim$18 per bunch crossing. To achieve high luminosity, there would be 3000 bunches in the machine with each bunch having 10^{11} protons. The luminosity of the LHC machine will be extremely high so that very rare interactions can be observed, but at the same time production of the background events will also be very high. The LHC will generate about a billion events in 1 s, of which $\sim$10 to 100 events are expected to be from Higgs decay.

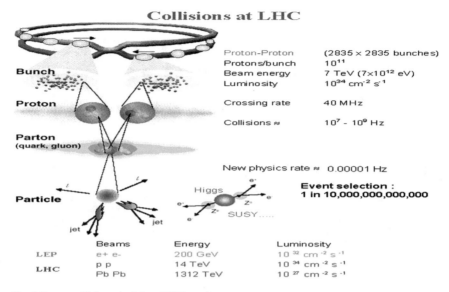

Fig. 1.3   p-p collision schedule at LHC accelerator

1.2.2   Physics at LHC

The fundamental physics program of LHC and CMS detector is to uncover and explore the physics of the electroweak symmetry breaking. Amongst others, some of them involve the following specific challenges:

1. SM Higgs boson search at masses above the maximum reach of Tevatron, i.e., ~115–1000 GeV.
2. Minimal Supersymmetric Model (MSSM) Higgs boson (h^0, H^0, A^0, $H^{\pm}$) searches up to masses of 2.5 TeV.
3. Search for new heavy gauge bosons ($W^{\pm}$, $Z^{\pm}$) up to masses of 4.5 TeV.
4. Search for supersymmetry (SUSY) partners of quarks and gluons-squark and gluino up to masses 2.5 TeV.
5. Search for composite structure of quarks and leptons.
6. Study of CP-violation.
7. Detailed studies of production and decays of top quark.
8. Test of electroweak (EW) gauge couplings-triple gauge boson vertices.
9. Search for Quark Gluon Plasma (QGP) in heavy ion collision.

One of the most important physics topic at CMS is the discovery potential of SM Higgs boson. In the search for SM Higgs boson, the LEP–II run at $\sqrt{s} = 195$–200 GeV extended the Higgs mass search up to $M_H = 95$–114 GeV

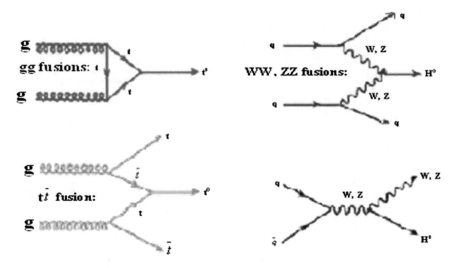

Fig. 1.4 Feynman diagram of Higgs production mechanism

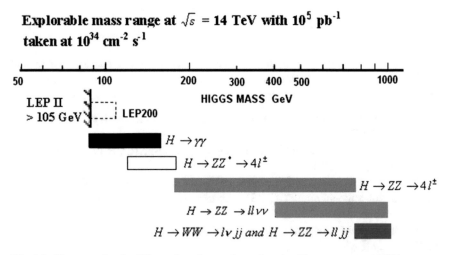

Fig. 1.5 Higgs searches in different detection modes and explorable mass range at LHC

[8]. Tevatron collider at FNAL may not be able to extend the Higgs mass reach beyond LEP-II. Thus, Higgs mass range of interest at LHC would be $M_H > 115$ GeV. Figure 1.4 shows the Feynman diagram of Higgs production mechanism [8] and Fig. 1.5 shows Higgs searches in different detection modes and explorable mass range at LHC.

1.3   The Compact Muon Solenoid (CMS) Experiment

The name Compact Muon Solenoid, well summarizes the main characteristics of the experimental apparatus: great emphasis has been put on building a highly efficient muon detection and measurement system, and the detector is based around the use of a single superconducting solenoid. CMS detector is based on a very high solenoidal field (~4 Tesla) and will identify and measure muons, electrons and photons over a wide momentum range.

Most detectors for particle physics are based around a magnet system, to facilitate the measurement of the momenta of charged particles. CMS will use a large superconducting solenoid, with a length of around 13 m and an inner diameter of about 6 m. The field strength will be 4 Tesla about 100,000 times that of the earth's magnetic field. The CMS, when working, will detect all sorts of interesting things including, if existing, the Higgs boson, the particle which may explain why matter has mass.

1.3.1   General Purpose CMS Detector

CMS will be one of the two general-purpose detectors at the LHC CERN. Figure 1.6 shows the 3-dimensional view of the CMS detector and Fig. 1.7 shows its superconducting solenoid magnet with a high field of 4 Tesla. The magnetic flux is returned through saturated iron return yoke which houses the barrel and muon chamber. The detector consists of an inner tracker with an embedded pixel detector, a crystal electromagnetic calorimeter, a copper-scintillator hadron calorimeter and a muon system made up of tracking chambers and special trigger chambers [3].

CMS is a large, technologically advanced detector made up of many layers. Each of these is designed to perform a specific task and together they will allow CMS to identify and precisely measure the energies of all the particles produced in the LHC p-p collisions. The layers of the CMS detector are arranged like a cylindrical onion around the collision point.

1.3.2   Sub-detectors in CMS

There are four sub-detectors in CMS: (1) Tracker Detector (Inner tracker detector), (2) Electromagnetic Calorimeter (ECAL), (3) Hadron Calorimeter (HCAL), (4) Muon Detector (Outer tracker detector). The CMS detector is roughly cylindrical and its sub-detectors are arranged in layers around the collision point, with different components identifying different particles. Figure 1.8 shows the general view of the detector in the form of layers used in the CMS experiment.

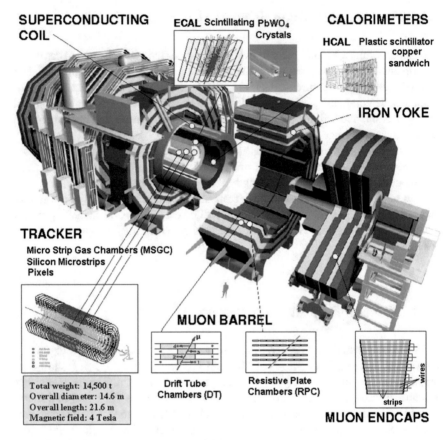

Fig. 1.6 3-D view of CMS detector at LHC CERN

1.3.2.1 Tracker Detector (Inner Tracker Detector)

A robust and versatile tracking detector is of utmost importance in CMS. A particle emerging from the p-p collisions in the centre of the CMS detector and traveling outwards will first encounter the inner tracker, which is a collection of three different detectors: pixels, gas microstrip chambers, and Si microstrip detectors. They will measure precisely the track, charge, momentum of the charged particles like electrons (e^-) and positrons (e^+), muons, positive pions (π^+), negative pions (π^-) etc. Neutral particles, i.e., photon (γ), neutron (n) and neutral pions (π^0) do not produce the track.

A large superconducting solenoid magnet with a high magnitude of field ~4 Tesla built in a CMS detector causes charge particle to follow spiral path in inner tracker helping to identify individual particle types. It provides high magnetic fields to bend these tracks so that particle momentum can be calculated from the amount of bending.

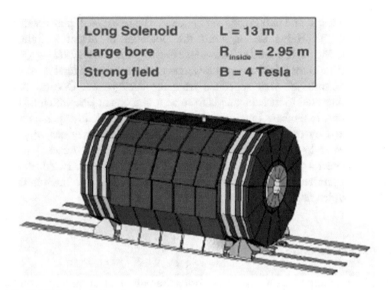

Long Solenoid	L = 13 m
Large bore	R_{inside} = 2.95 m
Strong field	B = 4 Tesla

Fig. 1.7 CMS Magnet Coil and return yoke

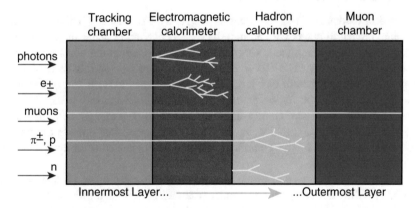

Fig. 1.8 General view of the detector in the form of layers used in the CMS experiment

1.3.2.2   Electromagnetic Calorimeter (ECAL)

The aim is to create calorimeter layer (ECAL) that measures the energies of $e^{\pm}$, γ with great precision. Outside the inner tracker, the next layer traps and identifies all $e^{\pm}$, γ and particle jet as they plough into dense material such as Lead Tungstate ($PbWO_4$). This material is interleaved with detectors to measure the energy that the particle loses as they come to halt. The CMS ECAL consists of one crystal barrel, two crystal endcaps, and two Preshower endcaps. The ECAL will play a vital role in the study of the Higgs sector, with the potential to detect a light Higgs (80 GeV/c^2 < M$_H$ < 130 GeV/c^2) through the two-photon decay ($H \rightarrow \gamma\gamma$) mode. The branching ratio for this channel is low, and consequently the total cross-section, and the signal must be well separated from the background. One of the major reducible background to this channel is from neutral pions in jets, which fake single isolated photons. In the barrel region these photons are sufficiently apart so that they can be detected by crystal calorimeter, but in the endcaps they are very closely spaced. In order to provide a better γ-π^0 separation in the forward region, a Preshower Detector (PSD) is placed in front of crystal endcaps. ECAL also helps in identifying particles such as neutral pions (π^0), which leave no track in the inner tracker, but which decay into two photons ($\pi^0 \rightarrow \gamma\gamma$).

1.3.2.3   Hadron Calorimeter (HCAL)

A third layer incorporates the iron that forms the outer part of the superconducting solenoid magnet. It stops the strongly interacting particles, i.e. hadrons like $\pi^\pm$, p etc. and measure their energies. Two kinds of particles, i.e., muons and neutrinos are likely to penetrate beyond the HCAL.

Barrel and Endcaps The hadron barrel (HB) and hadron endcap (HE) calorimeters are sampling calorimeters with 50 mm thick copper absorber plates interleaved with 4 mm thick scintillator sheets. Copper has been selected as the absorber material because of its density. The HB is constructed of two half-barrels each of 4.3 m length. The HE consists of two large structures, situated at each end of the barrel detector and within the region of high magnetic field. Because the barrel HCAL inside the coil is not sufficiently thick to contain all the energy of high energy showers, additional scintillation layers (HOB) are placed just outside the magnet coil. The full depth of the combined HB and HOB detectors is approximately 11 absorption lengths.

Forward: There are two hadronic forward (HF) calorimeters, one located at each end of the CMS detector, which complete the HCAL coverage to |η| = 5. The HF detectors are situated in a harsh radiation field and cannot be constructed of conventional scintillator and waveshifter materials. Instead, the HF is built of steel absorber plates; steel suffers less activation under irradiation than copper. Hadronic showers are sampled at various depths by radiation-resistant quartz fibers, of selected lengths, which are inserted into the absorber plates.

1.3.2.4   Muon Detector (Outer Detector)

Muons are charged particles, which will leave track further in muon chamber region and their momenta will be measured from the bending of their path in the magnetic field. Only neutrinos escape the apparatus without direct detection. But one can infer their existence using the principle of conservation of momentum and energy. The missing amount of energy is generally in the form of neutrinos (typically called missing energy).

1.4   Preshower Detector (PSD): Raison d'ê tre

Figure 1.9 shows the position of the Preshower in the ECAL. The Preshower is a disc shaped structure, required to measure the position of the 2 closely spaced photons, and is situated between the inner tracker and the ECAL. It is supported by some thin

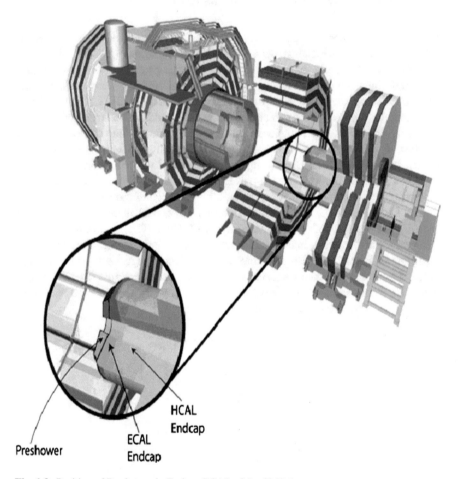

Fig. 1.9 Position of Preshower in Endcap ECAL of the CMS detector

aluminum (Al) struts at $\eta \sim 1.5$. Preshower will utilize more than 4300 Si sensors, totaling around 16 m^2 area. Each of the 32 strips on each Si sensor measures around 1.9 mm $\times$ 61 mm $\times$ 300 μm [9]. The Si sensors are produced by different manufacturers in Greece, India, Russia, and Taiwan on 4″ Si wafers.

The endcap region of CMS will experience a harsh radiation environment: the "hottest" part of the Preshower ($\eta \sim 2.6$) will receive around 6 Mrad and 2×10^{14} n cm^{-2} after ten years running. Under these conditions the Si detectors undergo "type inversion", and will require a high bias voltage to achieve full Charge Collection Efficiency (CCE).

1.4.1 Why Si Detectors Are Used in PSD?

As mentioned earlier, the granularity of the ECAL crystal in the endcap is not sufficient to distinguish energy deposit from a single photon or two closely spaced photons. A high-granularity photon-sensitive detector placed in front of the ECAL can discriminate single-photon energy deposits from double-photon ones, and will be able to 'throw away' some of the background events. In CMS, this function is performed by the Preshower detectors (one in each endcap). Neutral pions (π^0) are copiously created in p-p collisions and they decay mostly into 2 closely spaced photons. π^0 forms an important background to the Higgs decay (H $\rightarrow \gamma\gamma$) mode particularly in the endcap ($1.65 < |\eta| < 2.6$). Hence CMS will utilize a microstrip sensor as a PSD in the endcap region of ECAL. Its main function is to provide γ-π^0 separation [10]. Figure 1.10 shows the rapidity coverage of the PSD.

The CMS Preshower detector contains two orthogonal planes of Si microstrip sensors, each preceeded by two thin layers of Lead (Pb) absorbers (to initiate electromagnetic shower). Figure 1.11 shows the schematic layout of the Preshower Detector. Si detectors are also best suited to meet the challenges of high accuracy and efficient track measurements because of the high level of possible segmentation in strips. Their compactness and the possibility to integrate the front-end electronics on the same chip are also advantageous.

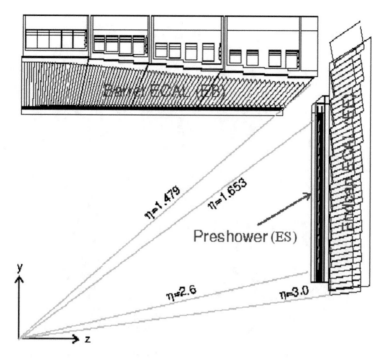

Fig. 1.10 Longitudinal view showing the Preshower

1.4.2   Expected Performance of the Preshower Detector to Single and Double Photon

Preshower detector is able to differentiate the single photon and closely spaced two photons produced in the p-p collisions. Figure 1.12 shows the shower detection in two orthogonal planes of Si microstrip detector in the endcap region of ECAL, Fig. 1.13(a) shows the transverse shower shape for single photon and Fig. 1.13(b) shows the same shape for two closely spaced photons from a π^0 decay. It can be seen that transverse shower shape is different for single photons and double photon.

Finally, the Higgs Boson (125–127 Gev/c^2) "God Particle" is detected on 4 July 2012 at CERN and that is reported by CMS and ATLAS experiments.

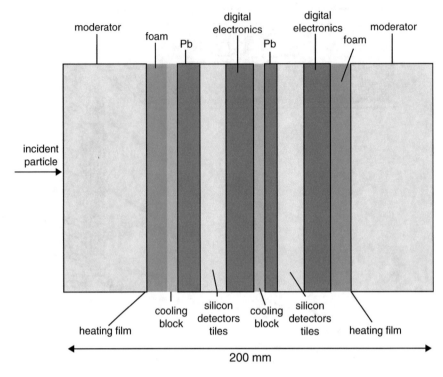

Fig. 1.11 Schematic layout of the Preshower Detector

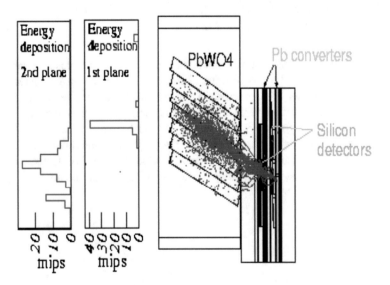

Fig. 1.12 Shower detection in two orthogonal planes of Si microstrip detector in the endcap region of ECAL

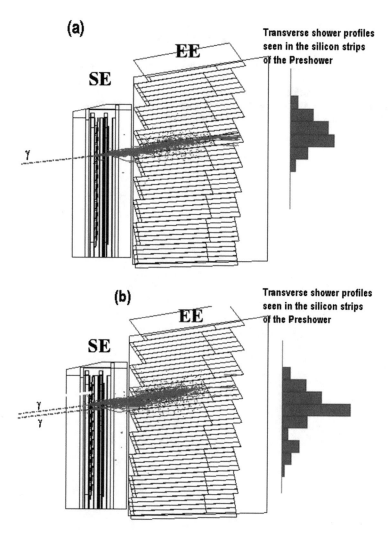

Fig. 1.13 (a) Transverse shower shape for single photon. (b) Transverse shower shape for two closely spaced photons from π^0 decay

References

1. Perkins, D.H.: Introduction to High Energy Physics, 2nd edn. Addison-Wesley, Reading, MA (1982)
2. Dimopoulos, S., Lindner, M.: Proc. of LHC Workshop Auchen, CERN/90–10 (1990)
3. The CMS Collaboration: Technical Proposal, CERN/LHCC 94–38 (1994)
4. Kirti Ranjan Ph. D. Thesis. Large Transverse momentum (P_T) particle production in p-p collision at LHC energies. Department of Physics and Astrophysics, University of Delhi, India (2002)

5. Singh, J.B.: Pramana – J. Phys. **54**(4) (April 2000)
6. http://www.quark.lu.se/~atlas/thesis/egede/thesis-node9.html
7. http://cmsinfo.cern.ch
8. Beenakker, W., Hopker, R., Spira, M., Zerwas, P.: Nucl. Phys. B492, 51, Spira, hep-ph/9705337 (1997)
9. http://cmsdoc.cern.ch/cms/ECAL/preshower
10. Kyriakis, A. et al.: An Artificial Neural Net Approach to Photon – Pi-zero Discrimination using the CMS Endcap Preshower. CMS NOTE-088/1998

Chapter 2
Physics and Technology of Si Detectors

In the early 1960s, the first semiconductor detector was used in the electronics industries. After the first implementation of planar technology in 1980 by J. Kemmer [1–3], semiconductors were quickly understood to give detectors of extraordinarily high performance in High Energy Physics (HEP) experiments. Semiconductor detectors have been used in modern HEP experiments in the form of pixel detectors, strip/microstrip detectors. They are popular due to their unmatched energy and spatial resolution and excellent response time. These detectors are manufactured mainly of silicon (Si), traditionally on high-resistivity single crystal float-zone material. Recent progress in micro technology now allows reliable large-scale production of detectors of sophisticated designs at acceptable cost.

Advantages of Si Detectors Si detectors are more suitable for particle identification and tracking in various ongoing and future HEP experiments because of their definite advantages over other detectors available for the same purpose, which includes:

- fast response of the order of few ns.
- possibility of high level of segmentation-strips, microstrips, pixels etc.
- excellent energy and position resolution.
- simplicity of operation at room temperature.
- small size.
- linearity in response.
- compactness.
- high stability.

Use of Si as a substrate material in detector technology enables batch fabrication with very good uniformity and low cost of production.

© Springer Nature Switzerland AG 2019   17
A. K. Srivastava, *Si Detectors and Characterization for HEP and Photon Science Experiment*, https://doi.org/10.1007/978-3-030-19531-1_2

2.1   Characteristics of Si Microstrip Detectors

The fundamental structure of Si detectors is basically one sided abrupt p^+-n ($N_a \gg N_d$) junction. It is usually operated in Reverse Biased (RB) mode which provides necessary condition for the detection of ionizing radiation. In RB mode, signal current pulse is larger than noise due to its leakage current and therefore, the signal can be measured with good energy resolution. Blocking electrodes are usually employed to reduce the magnitude of the leakage current through the n-bulk. The leakage current in reverse biased Si detector can be reduced to sufficiently small values to allow the detection of the added current pulse created by the electron (e)–hole (h) pairs produced along the track of an ionizing particle. Figure 2.1 shows the schematic of Si microstrip detector.

The first Si micro-strip detectors were created in the early 1980's, and they are frequently are now used as front line detectors for most modern particle accelerator experiments. The structure of Si microstrip detector is very similar to the old semiconductor devices. They still have an n-side and a p-side, but instead of having a layer of doped Si on each surface, these are strips of doped Si running parallel to each other along the length of the detector.

The high resolution of the detectors is needed to accurately measure the track and vertices of particles, particularly the secondary vertices produced in the decay of heavy quarks, created in a sub-atomic particle collision.

The position accuracy of Si detectors is achieved by dividing the p-n diode into fine parallel strips that act as individual independent diodes. Si detector segmented into number of long narrow strips (p^+) is known as Si microstrip detector (shown in Fig. 2.1). One of the important geometrical parameter of a Si microstrip structure is

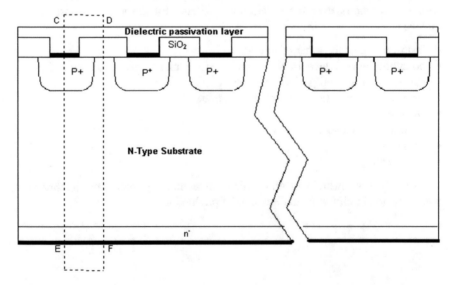

Fig. 2.1 Cross-sectional view of Si microstrip detector

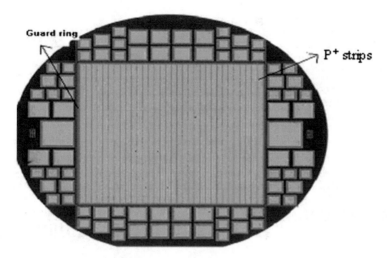

Fig. 2.2 Top sectional view of 4″ BEL Si microstrip detector

the pitch, with typical pitch values in the range of ten to hundred micrometers depending on the application. Si microstrip detectors can be built as single-or double-sided devices. The readout is normally done in channels connecting several strips.

Various type of Si strip detectors are being used in HEP experiments:

- Single sided and double sided strip detectors (DC coupled, 2D Position sensing).
- Pixel detectors (suitable for imaging applications).
- Si microstrip detectors (AC coupled, single or double sided).
- Si drift detectors (high energy and position resolution, suitable for imaging applications).
- Monolithic active pixel detectors.
- Single element detectors with high energy resolution/large sensitive area.

The fabrication of single sided Si microstrip detectors for Preshower of CMS have been done at Bharat Electronics Limited (BEL), Bengaluru, India. The dimension of Si detector was 63×63 mm^2 to take advantage of the whole surface of the available 4″ wafer [4]. Figure 2.2 shows the top view of 4″ BEL Si microstrip detector.

2.2   Semiconductor Properties and p-n Junction

Semiconductor particle detectors use Si ($_{14}$Si^{28}) as an indirect semiconductor primarily in a single crystal formation. Table 2.1 shows some properties of an intrinsic Si crystal.

Table 2.1 Some properties of a pure Si crystal

Properties of pure Si	Representation	Value (in units)
Intrinsic carrier concentration (n_i) at 300 °K	$n_i = n_e = n_h$	1.45×10^{10} cm^{-3}
Temperature dependence	σ	$e^{-Eg/2kT}$
Forbidden gap (at 25 °C)	E_g	1.1 eV
Dielectric constant	ε_{Si}	11.7
Crystal structure	Diamond type	–
Lattice constant	a	5.43 Å
Density	D	2.33 gcm^{-3}
Electron mobility (at 300 K)	μ_e	1350 cm^2 V^{-1} s^{-1}
Hole mobility (at 300 K)	μ_h	480 cm^2 V^{-1} s^{-1}
Intrinsic resistivity	ρ_0	2×10^5 Ω cm
Energy require to create an e-h pairs	W	3.66 eV
Melting point	–	1415 °C

The electrical conductivity (σ) of the semiconductor (Si, Germanium (Ge) etc.) lies in between that of the metals and insulators. Electrical conductivity is generally given by the energy band representation.

The gap between the valence and the conduction band is known as forbidden gap. At absolute zero (0 K), no free electrons are available for electrical conduction and hence semiconductors behave as an insulators. However, at higher temperatures, the electrons can be thermally excited out of a covalent bond and become free to participate in electrical conduction. This leaves electron vacancies in the covalent bond referred to as "hole". For intrinsic Si, Fermi level (E_F) lies midway in between the conduction band and valence band. When small amount of trivalent impurity (Boron (B), Aluminum (Al) etc.) is added to Si then it is known as p-type Si and when pentavalent impurity (Phosphorous (P), Arsenic (As) etc.) is added to Si then it is known as n-type Si.

The movement of holes also contributes to the electrical conductivity of the semiconductor material. The value of the electrical conductivity (σ), of the semiconductor material is related to the charge carrier densities of free electrons (e) and holes (h) and their mobility as follows [5],

$$\sigma = q(\mu_e n_e + \mu_h n_h). \qquad (2.1)$$

The resistivity of the semiconductor material (ρ) is equal to the inverse of the electrical conductivity, and is given by the following expression;

$$\rho = \frac{1}{q(\mu_e n_e + \mu_h n_h)}, \qquad (2.2)$$

where μ_e, μ_h represent electron and hole mobility respectively and n_e, n_h represent the respective electron and hole density.

Photons with energies greater than the band gap energy of Si can excite an electron into the conduction band and leaving a hole in the valence band (Si has forbidden gap of 1.1 eV). The ability of Si to collect these created charges allows its use as a particle detector. However, for this purpose, almost all the free charge carriers are required to be removed from the Si bulk.

2.2.1 Electrical Characteristics of p-n Junction

If we bring p and n-type semiconductor materials into contact to form a p-n junction, carrier diffusion takes place as holes from the p side diffuse into the n side, and electrons diffuse from n to p side.

The diffusion current, however, cannot build up indefinitely because an opposing internal electric field is created at the p-n junction. To clarify, electrons diffusing from n to p leave behind uncompensated donor ions (N_D^+) in the n material, and holes leaving the p region leave behind uncompensated acceptors (N_A^-). Thus, a region of positive space charge near the p side and negative charge near the p side would develop. Therefore, the resulting electric field creates a drift component of current from n to p opposing the diffusion current.

The electric field and potential distribution in the depletion region can be calculated by solving one dimensional Poisson equation as given by the following expression [6];

$$\frac{d^2V}{dx^2} = -\frac{\rho}{\varepsilon_{Si}} = -\frac{\rho}{\varepsilon_r\varepsilon_0},\qquad(2.3)$$

where dielectric constant for Si, $\varepsilon_{si} = \varepsilon_r\varepsilon_0$. The value of relative permittivity (ε_r) for Si is 11.7 and value of permittivity in vacuum (ε_0) is 8.85×10^{-12} Fm^{-1}. Figure 2.3 shows the energy band diagram after merging the p-type and n-type material.

The width of the region with no free charge carriers is known as the depletion region or space charge region (SCR) and this can be determined using the following expression;

$$W = \left[\frac{2V_0\varepsilon_{Si}}{q}\left(\frac{1}{N_A} + \frac{1}{N_D}\right)\right]^{1/2},\qquad(2.4)$$

where ε_{si}, q are constants and N$_A$, N$_D$ represent concentration of acceptors and donors. V$_o$ stands for the potential difference that builds up between the n-type and p-type layers.

Equation (2.4) indicates that the width of the depletion region can be altered by applying external voltage across the junction. A positive voltage applied on the p side and a negative voltage on the n side (Forward Biasing (FB)) would oppose the built up internal voltage, as a result of which the depletion region would be reduced

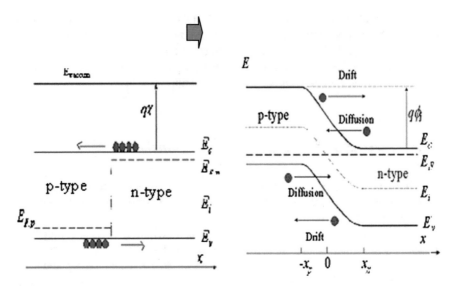

Fig. 2.3 Energy band diagram after merging the n-type and p-type regions

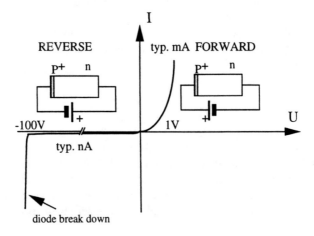

Fig. 2.4 Current-voltage characteristics of a diode under forward and reverse biasing

and the current between the terminals grows exponentially. Similarly, when negative voltage is applied on the p-side and positive voltage on the n-side (Reverse Biasing (RB)), the depletion region would grow. The effect of biasing of p-n junction on the resulting current from the junction is depicted in Fig. 2.4.

A p-n junction diode is normally fabricated by ion implantation of p-type dopant into the surface of an n-type Si wafer. The density of p-type doping centers (acceptor impurity concentration N_A^-) in the surface layers is many times higher than the density of n-type centers (donor impurity concentration N_D^+) in the n-type bulk Si crystal.

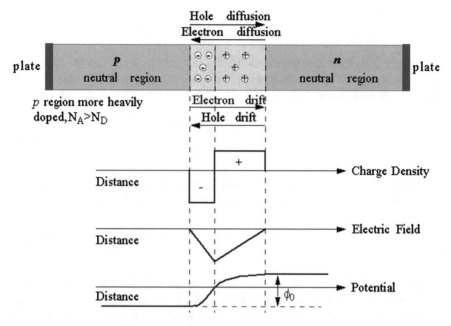

Fig. 2.5 Charge distribution, electric field and electrostatic potential plot in one-sided abrupt p^+-n junction

$$N_A \gg N_D \tag{2.5}$$

The sudden change in the doping concentration at the p-n junction over a few atomic layers has earned the name 'one sided abrupt junction' for the p-n junction. This simple junction finds widespread application for the detection of the ionizing radiations [7]. Figure 2.5 shows the charge distribution, electric field and electrostatic potential plot of one-sided abrupt p^+-n junction.

For p^+-n junction, the width of the depletion region (W) is given by,

$$W = \sqrt{\frac{2\varepsilon_{Si}V_0}{qN_D}}, \tag{2.6}$$

and built-in-potential (V_0) is given by:

$$V_0 = \frac{KT}{q} \ln \frac{N_A N_D}{n_i^2}. \tag{2.7}$$

Also,

$$V_0 = \frac{1}{2} E_{max} W, \tag{2.8}$$

where E_{max} is the maximum electric field at the junction. As discussed earlier, p^+-n junction detector is usually operated in RB mode. For an applied bias voltage (V), W in RB mode is given by,

$$W = \sqrt{\frac{2\varepsilon_{SI}(V_0 + V)}{qN_D}}, \tag{2.9}$$

The voltage necessary to extend the depletion region over an entire detector is known as a full depletion voltage (V_{FD}) and calculated as follows;

$$V_{FD} = \frac{qN_d W_N{}^2}{2\varepsilon_{Si}}, \tag{2.10}$$

where W_N is the thickness of the detector. As stated earlier, the application of Si as a particle detector depends upon the presence of a minimum number of free charge carriers. This is achieved by applying the necessary voltage to a reverse biased p-n junction. In fact, in order to avoid losses in charge collection, the Si detectors are over biased. There is, however, one key limiting factor to applying maximum reverse bias voltage: the breakdown phenomenon.

The electric in a full depleted detector at a distance $x = W_N$ and as a function of depth (x) can be calculated by the following expression:

$$E(W_N) = \frac{V - V_{FD}}{W_N} \tag{2.11}$$

$$E(x) = \frac{2V_{FD}}{W_N}\left(1 - \frac{x}{W_N}\right) + \frac{V - V_{FD}}{W_N} \tag{2.12}$$

Electric field versus depth of the (a) non-depleted and (b) over depleted detector is shown in Fig. 2.6.

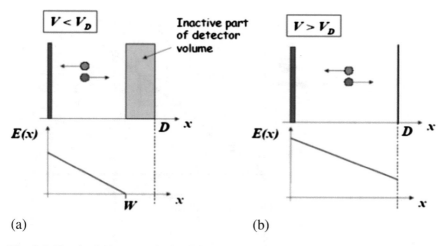

Fig. 2.6 Electric field versus depth of the (**a**) non-depleted, and (**b**) over depleted detector (D-detector thickness and V_D is full depletion voltage)

2.3   Physics of Reverse Biased p⁺-n Junction Microstrip Detector

The physics of reverse biased p⁺-n junction is described below.

2.3.1   The Reverse Leakage Current

In reverse biased p-n junctions, there is a paucity of charge carriers in the depletion region and e-h pair once generated, get separated under the influence of electric field and hence their probability of recombination is diminished. Thus, generation mechanism is chiefly responsible for the current-flow in the reverse biased silicon detectors. The reverse leakage current is the sum of the several components: reverse leakage current = saturation current + generation current + surface current. Saturation current flows due to the minority charge carriers and generation current is related to the existence of the generation-recombination centers. To avoid surface current, surface passivation is necessary in p⁺-n junction devices. Diffusion current from generation at edge of the depletion region is negligible for a fully depleted detector.

The generation current is given by the

$$I_{gen} = \frac{q n_i A W_N}{\tau_{g,eff}} \tag{2.13}$$

where n_i is the intrinsic carrier density at a given temperature, A is the active area of the sensor, and $\tau_{g,eff}$ is the effective generation life time of the charge carrier in the presence of the generation-recombination centers (deep level traps).

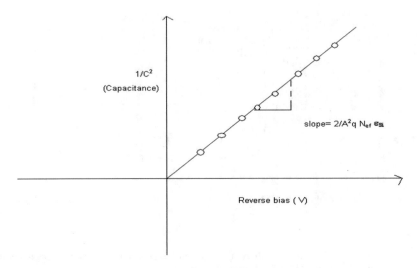

Fig. 2.7 The standard way of plotting the CV data of one-sided abrupt p$^+$-n junction

2.3.2 Capacitance

Capacitance is due to the movement of charges in the junction. A reverse biased p$^+$-n diode can be considered as a parallel plate capacitor. The depletion layer capacitance (C_{dep}) of the parallel plate capacitor is given by [6];

$$C_{dep} = \frac{\varepsilon_{Si}A}{W} = A\left[\frac{q\varepsilon_{Si}N_{eff}}{2}\right]^{1/2}\frac{1}{\sqrt{V}}, \qquad (2.14)$$

where V is the applied reverse bias. Here, built-in-voltage (V_0) is neglected, because $V \gg V_0$ is assumed. The capacitance decreases proportionally to the square root of the voltage until the depletion region extends to the full depletion of the diode.

From Eq. (2.11) we get,

$$\frac{1}{C^2} = \frac{W^2}{\varepsilon_{Si}A} = \frac{2V}{A^2 q\varepsilon_{Si}N_{eff}}. \qquad (2.15)$$

Equation (2.15) shows a linear relationship between 1/C^2 versus V. Figure 2.7 shows the plot of 1/C^2 versus V for the p$^+$-n junction. From the slope of the line, we can get the effective doping concentration.

The bias voltage at which full depletion is attained is called full depletion voltage (V_{fd}). For bias voltage higher than V_{fd} the capacitance is constant and its value geometrical capacitance (C_{fd}.) can be considered to be that of the parallel plate capacitor.

2.3.3 Noise

The degradation in the signal delivered by the detector is known as noise. The energy resolution of semiconductor detector systems is determined largely by electronic noise. Since the energy deposited in the detector translates directly into charge. It is convenient to express the electronic noise as an equivalent noise charge (ENC). Basically, it depends upon the detector capacitance, reverse current, electronics design and shaping time. Lower capacitance implies the lower noise whereas faster electronics makes the noise contribution from reverse current less significant. To achieve a high performance of Si microstrip detectors for the CMS experiment at LHC CERN, high signal/noise (S/N) ratio is also one of its important requirements [8].

2.3.4 Charge Collection Efficiency (CCE)

The charge collected by the electrodes of the Si detector is known as charge collection. The charge deposited by a Minimum Ionizing Particle (MIP), Q_{dep} or (Q_0), is given by the following expression;

$$Q_{dep} = \frac{dE}{dx} W_N,$$ (2.16)

where W_N is the width of the detector and dE/dx is the energy loss per unit length of detector traversed. If the charge carrier has a mean free path λ, then the charge collected Q_{col} or (Q) is given by,

$$Q_{col} = Q_{dep} \frac{\lambda}{W_N}.$$ (2.17)

Hence,

$$\lambda = \frac{Q_{col}}{dE/dx}.$$ (2.18)

The charge collection efficiency of Si detector is given by

$$CCE = \frac{Q(t)}{Q_0}$$ (2.19)

where collected charge (Q) will be corrected for the trapping time of the charge carrier inside irradiated Si detector and can be write $Q(t) = Qe^{\frac{-t}{\tau_{eff}}}$, where τ_{eff} is the effective trapping time constant.

2.3.5   Avalanche Breakdown Voltage in Reverse Biased p⁺-n Junction

An important mechanism for the occurrence of breakdown in reverse biased p⁺-n junction is avalanche multiplication or the impact ionization of host atoms by charge carriers in high electric field [5–7]. To illustrate this phenomenon, imagine that the electric field in the depletion region is large enough so that an electron entering from the p side may be accelerated to high enough kinetic energy to cause an ionizing collision with the Si/dopant crystal lattice. As a result of this interaction, the original electron and the generated electron would both be swept to the n side, while the generated hole would be swept to the p side. On their paths, however, the resulting carriers also have the chance of creating a new electron-hole pair, and each of those can also create an electron-hole pair, and so forth.

To approximate the degree of carrier multiplication, assume that a carrier of either type has a probability (P) of having an ionizing collision and that there are n_{in} number of electrons entering from the p side. This will give us the total number of electrons out of the region (n_{out}) after many collisions given by,

$$n_{out} = n_{in}\left(1 + P + P^2 + P^3 + \ldots\right). \tag{2.20}$$

The electron multiplication factor (M_n) is defined as the ratio of n_{out} to n_{in} [9],

$$M_n = \frac{n_{out}}{n_{in}} = 1 + P + P^2 + P^3 + \ldots = \frac{1}{1 - P}. \tag{2.21}$$

Physically, we would expect the ionization probability to increase with the applied voltage. Measurements of carrier multiplication support this prediction by an empirical relation [9];

$$M = \frac{1}{1 - \left(V/V_{BD}\right)^n}, \tag{2.22}$$

where V_{BD} is the breakdown voltage, the critical voltage at which point the reverse current through the junction increases sharply and relatively large currents can flow with little further increase in voltage. The value of the breakdown voltage can be further quantified through the following expression [9];

$$V_{BD} = \frac{E_c^2 \varepsilon_{Si}}{2qN_{eff}}, \tag{2.23}$$

where, N_{eff} represents the effective carrier concentration and E_c is the critical electric field.

2.3.6 Techniques to Improve V_{BD} of p$^+$-n Junctions

Various type of commonly used p$^+$-n junctions have been used in the design of Si microstrip detectors. Figure 2.8(a–d) shows the different types of p$^+$-n junctions. It is known that the real devices with curved p-n junctions (Fig. 2.8b) have significantly lower V_{BD} than the plane-parallel idealization (Fig. 2.7) [6].

Breakdown Voltage (V_{BD}) is a crucial factor in the long-term operation of the Si detector. It depends on the detector design and on fabrication technology. Electrode geometry, number and size of guard ring, oxide quality and junction fabrication can also influence V_{BD} [10]. The radiation damage increases the reverse leakage current and requires the detector to operate at high voltage which increases the chance of the occurrence of the breakdown. Therefore, the detector breakdown is one of the limits to the operation of Si detector after irradiation. This problem of irradiated detector can be solved by design solutions [10].

Breakdown voltage of p$^+$-n junction strip detector can be increased by incorporating the guard rings or Field Plate (FP) also known as metal overhangs which is the extension of metal on insulator SiO$_2$. The adoption of "overhanging" metal contacts helps in distributing the electric field, reducing corner effects and thus minimizing breakdown risks [11]. The guard ring is used to define the active depletion region and to prevent the leakage current from being absorbed by the edge strips, which would depreciate their performance. Multiple guard ring structures can increase the breakdown limit of the Si microstrip detectors, provided the guard ring spacing is carefully optimized. Equal values of surface electric field at the main junction and the guard rings can serve as the criterion for optimization of multi-guard structures [12].

For normal cylindrical curvature junction (see in Figs. 2.9 and 2.8b), electric field lines emanate from the charge Q_C and terminate on negative charge in the curved portion of p$^+$-edge depletion region The electric flux per unit area crossing the curved portion of the junction is more than that at the planar portion of the junction, and there is a field crowding at the curved portion. This field crowding limits the V_{BD} of normal cylindrical junction [13]. For normal cylindrical curvature p$^+$-n junction equipped with field plate, two peaks in the electric field distribution ("E_{max1}" and "E_{max2}") are shown in Figs. 2.10 and 2.8c which increases the breakdown voltage of Si detector.

In all of these p$^+$-n junctions, Si-SiO$_2$ interface is very important. Charges in the oxide layer can have a dramatic effect on device characteristics. A charge at the interface can introduce a charge of the opposite polarity in the underlying Si and results in the contraction of the depletion region resulting in the premature breakdown of the device [14]. Charge situated at the Si-SiO$_2$ interface is shown in Fig. 2.11. Located at the Si-SiO$_2$ interface, interface–trapped charge (Q_{it}) have energy states in the Si-forbidden gap and can interact electrically with the underlying Si. The fixed oxide charge (Q_f) usually positive and it is located in the oxide within approximately 30 Å of the Si-SiO$_2$ interface. The interface oxide charge (Q_f) is ~3 × 10^{11} cm^{-2} for non-radiated high quality SiO$_2$ grown on Si with a <111>

Fig. 2.8 Different types of
p⁺-n junction (**a**) parallel
planar (**b**) normal cylindrical
curvature junction (**c**) p⁺-n
junction equipped with field
plate (FP) (**d**) p⁺-n junction
with guard ring

(a)

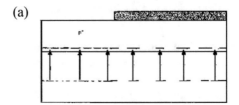

(b)

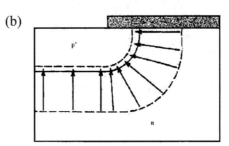

(c)

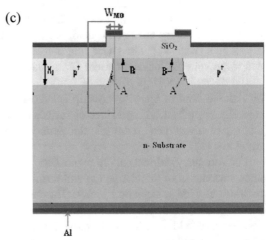

(d)

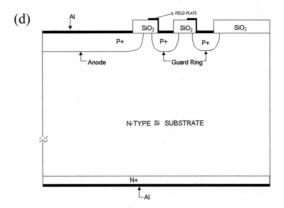

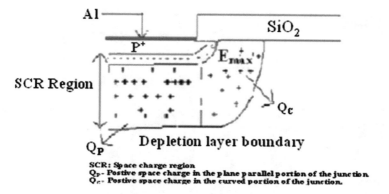

Fig. 2.9 Cross-section of normal cylindrical curvature junction

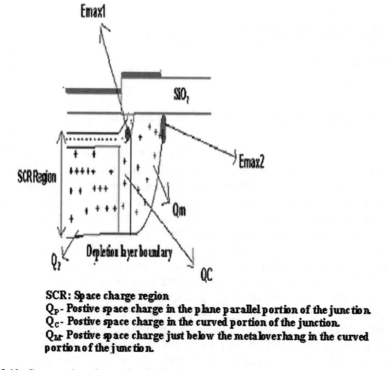

Fig. 2.10 Cross-section of normal cylindrical curvature junction with field plate

orientation. Under ionizing radiation, it rises up quickly and saturates at about 1×10^{12} cm^{-2}.

The oxide trapped charge (Q_{ot}) may be positive or negative near metal layer or in oxide bulk. Mobile ionic charge is usually due to alkali impurities such as sodium (Na⁺), potassium (K⁺), as well as to negative ions and heavy metals.

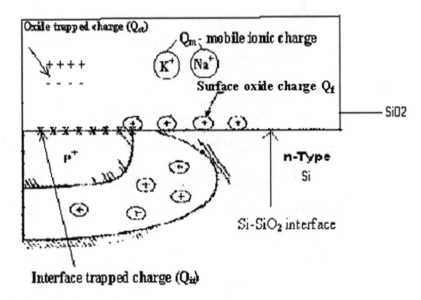

Fig. 2.11 Si-SiO$_2$ interface

2.4 Principle of Operation of p$^+$-n Si Microstrip Detector

Figure 2.12 shows the schematic representation of operation of single sided p$^+$-n Si microstrip detector (AC coupled). If an ionizing charged particle penetrates the detector it produces e-h pairs along its track, the number being proportional to the energy loss. An externally applied electric field (caused by the depletion voltage) separates the e-h pairs before they recombine; electrons drift towards the n$^+$ (anode), holes to the p$^+$ (cathode). The collected charge produces a current pulse on the electrode, whose integral equals the total charge generated by the incident charged particle, i.e., a measure of the deposited energy. The signal is picked up as current pulse. The readout goes through a charge-sensitive preamplifier, followed by a shaping amplifier. Amount of charge produced is dependent on the width of the depletion region and also directly proportional to Si microstrip detector thickness. For 300 micron thick Si detector, 30000 e-h pairs are produced (equivalent to 5fC), which is the charge produced by one ionizing particle, known as MIP. The collected charge can be used for calorimetric measurement. In Fig. 2.12, the n-bulk serves as the detection volume, while the n$^+$ back plane connection is used as an ohmic contact to the n-type material. The p$^+$ implant is used to deplete the N bulk and acts as a measuring electrode. The aluminum readout line serves as a contact to the input of an amplifier (referred to as a DC coupling contact). However, there would be a problem here because when the aluminum line is in direct ohmic contact with the strip implant, the constant detector leakage current would flow into the amplifier. This can result in serious operation problems by saturating the charge amplifier. A "bias

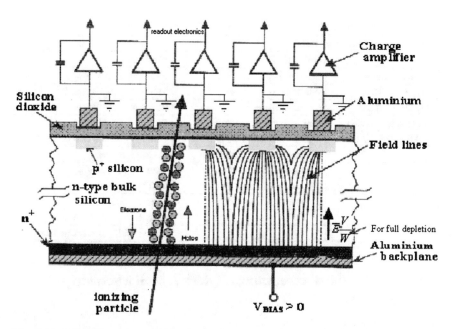

Fig. 2.12 Schematic representation of operation of single sided p$^+$-n Si microstrip detector (AC coupled)

line" present at the edge of the detector connects the individual strips through individual resistors to a fixed potential. The resistance for these bias resistors has to be high enough to separate the strips from one another. The resistor values also have to be constant since large variations in resistor values can lead to variation in the applied strip depletion voltage. Normally, there are two methods of coupling the detector: (a) DC coupling & (b) AC coupling. AC Coupled devices are preferred in order to avoid draining bulk current into electronics. Usually detectors are built with AC coupling, but the cost is higher than the DC coupled devices.

The following analytical expression used for the calculation of the time needed for charge carriers to transverse the entire detector,

$$v(x) = \mu E(x)$$
$$t(x_1, x_2) = \int_{x_1}^{x_2} dx/v_x \tag{2.24}$$

$$t_{drift} = \frac{W^2_N}{2\mu_i V} \ln\left(1 + \frac{2V}{V - V_{FD}}\right) \tag{2.25}$$

Table 2.2 Preliminary specifications/acceptance criterion of the Si sensors

Specifications (PRR-July 2001)		
Wafer	Geometry/Design	Electrical
4″ Float-zone Si	Length: 63.0 ± 0.2 mm	**Full depletion Voltage**
N-type <111>	Width: 63.0 ± 0.1 mm	$55\ V \leq V_L\ (0.32/t(mm))^2 \leq 150\ V$
t = 320 ± 20 μm	No. of Strips = 32	**Breakdown Voltage**
Single-side polished	Strip-pitch = 1.90 mm	Cat. 1: $V_L \geq 300\ V$
p~ 4.0 ± 0.5 kΩ cm	p^+ strip-width = 1.78 mm	Cat. 2: $V_L \geq 500\ V$ (~25%)
	p^+ strip-length = 60.82 mm	**Leakage current**
	A1 strip-width = 1.8 mm	**Total**: <5 μA at V_{FD}
	(MO = 10 μm on both sides)	<100 μA at 300 V
	n^+ layer thickness > 2.5 μm	**Strip-by-strip**
		Max. 1 with I > 1 μA at V_{FD}
		Max. 1 with I > 5 μA at 300 V

2.5 Specifications/Acceptance-Criterion of Si Microstrip Detectors

Specifications/acceptance-criterion provided by PSD group at CERN are listed in Table 2.2 [15, 16]. $V_{BD} > 500$ Volt is required for Si microstrip detectors to be placed in the central ring which is close to the beam pipe and where the received expected radiation flux is maximum.

Detector specifications are very stringent as these sensors are to be operated in intensely high radiation background of neutrons (2×10^{14} cm^{-2}) and gamma (10 Mrad.) etc. for a long period of ten years.

2.6 Fabrication of Si Microstrip Detector at BEL, India

J. Kemmer [3] introduced planar technology for the fabrication of the Si detector. The heart of any semiconductor manufacturing business is the "foundry," where the detectors/integrated circuits (IC's) are manufactured on the n-type/p-type Si wafer.

Geometrical specifications for the single sided Si microstrip detector have been provided by PSD group at CERN [16]. The fabrication steps used at BEL to fabricate the single sided Si microstrip detectors in an environmentally controlled clean room (class 100) are listed below:

1. Initial oxidation
2. p^+ lithography
3. Screen oxidation
4. Re-expose p^+ mask

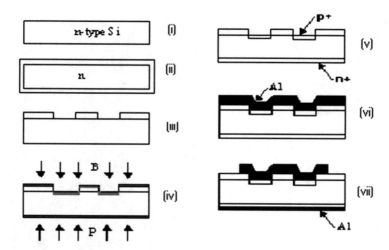

Fig. 2.13 Main steps in the fabrication of the detectors on n-type Si wafer

5. Implantation of boron for p$^+$ strips and guard rings
6. n$^+$ implant at the backside
7. Implant anneal and redistribution
8. Contact lithography
9. Front metallization
10. Metal lithography
11. Metal sintering at 450 °C
12. Final passivation

In total there are four masks including the final passivation. Figure 2.13 shows the main steps in the fabrication of the detectors on n-type Si wafer. Figure 2.14 shows the layout and cross-section of the Si microstrip detectors fabricated at BEL, India and complete 63 × 63 mm^2 CMS Preshower Si detector (with front-end electronics). Typically, it takes ~10–25 days to complete one batch of Si sensors. The thermal oxidation, masking, etching and doping steps are repeated many times until the last "front end" layer is completed. After the last metal is patterned, a final insulating layer (for final passivation) is deposited to protect the detectors from damage and chemical contamination. Openings are etched in this film to allow access to the top metal later by electrical probes and subsequent wire bonds. At present these detectors are in production phase at BEL.

Fig. 2.14 (**a**) Layout of the Si microstrip Preshower detectors (**b**) Cross-section of the Si microstrip Preshower detectors, and (**c**) Complete 63×63 mm^2 CMS Preshower Si detector (with front-end electronics)

(a)

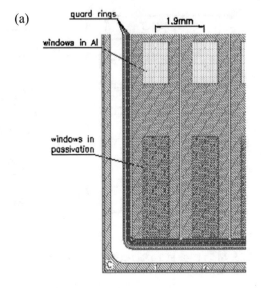

(b)

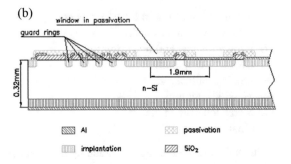

(c)

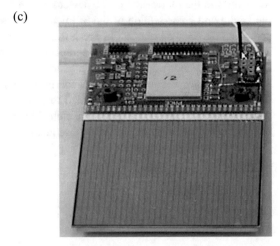

References

1. http://rkb.home.cern.ch/rkb/PH14PP/node206.html#smit80
2. http://rkb.home.cern.ch/rkb/PH14PP/node206.html#kemm80
3. Kemmer, J.: Nucl. Instr. Methods A. **169**, 499 (1980)
4. Apollo Go, Anna Peisert. Measurement of capacitance on Preshower detectors. CMS IN 2000/041
5. Sze, S.M.: Physics of Semiconductor Devices. Wiley, New York (1981)
6. Gandhi, S.K.: Semiconductor Power Devices. Wiley, New York (1977)
7. Grove, A.S.: Physics and Technology of Semiconductor Devices. Wiley, New York (1987)
8. Chatterji, S., Bhardwaj, A., Ranjan, K., Namrata, Srivastava, A.K., Shivpuri, R.K.: Solid State Electron. **47**(9), 1491 (2003)
9. Asmamaw, Z.: Electrical Engineering, Brown University, project report at Fermilab "Breakdown Voltage Measurements of Si Microstrip Detectors" (2001)
10. Bhardwaj, A.: Ph.D. thesis "Some aspects of hadron–hadron collision at the Large Hadron Collider – CERN". Department of Physics and Astrophysics, University of Delhi, India
11. Ranjan, K., Bhardwaj, A., Namrata, Chatterji, S., Srivastava, A.K., Shivpuri, R.K.: Semicond. Sci. Technol. **16**, 635 (2001)
12. Bharadwaj, A., Ranjan, K., Namrata, Chatterji, S., Srivastava, A.K., Shivpuri, R.K.: Semicond. Sci. Technol. **16**, 849 (2001)
13. Srivastava, A.K., Bhardwaj, A., Namrata, Chatterji, S., Shivpuri, R.K.: Semicond. Sci. Technol. **17**, 427 (2002)
14. Sze, S.M.: VLSI Technology. McGraw-Hill Book Company, Singapore (1988)
15. Piesert, A., Zamiatin, N.: CMS NOTE 2000–016 (2000)
16. Piesert, A.: Pre Shower Si PRR meeting CERN, Geneva (July 2001)

Chapter 3
Performance of MCz Si Material for $p^+n^-n^+$ and $n^+p^-p^+$ Si Sensor Design: Status and Development for HL-LHC

3.1 Introduction

LHC (Large Hadron Collider) at CERN is a one of the prestigious High-Energy Physics (HEP) experiment in World where several fundamental questions about Universe can be addressed. LHC (pp Collider, 14 TeV, 25 ns bunch spacing) will be upgraded to HL-LHC where the luminosity increases of up to ten times i.e., 10^{35}/cm^2 s [1] and as per our new HL-LHC TD report, LHC upgrade will bring the luminosity to about $5–7 \times 10^{34}$ cm^{-2} s^{-1} in 2026, with a goal of an integrated luminosity of 3000 fb^{-1} by the end of 2037. This High Luminosity LHC scenario, HL-LHC, will require a preparation program of the LHC detectors known as Phase-2 Upgrade. The radiation damage effects in the Si sensors at HL-LHC will be more challenging to cope with such hostile radiation environment therefore the Compact Muon Solenoid (CMS) experiments will require a new CMS tracking detectors for the HL-LHC. The active area covered by Si sensors (pixels, strixels, and strips) in the new CMS inner tracker detector region will be an order of 190 m^2.

Close to the p-p interaction point at HL-LHC (radius (R) < 20 cm), the following particles (e, p, n, π) will be produced from hadron-hadron collisions and pions will be dominant near to collision point (R < 4 cm) and pixel sensors will be used for the fluences up to 2.5×10^{16}/cm^2. The Si strixel (short strips of 2.5 cm or 5 cm long) sensors has been proposed for the outer strip region (20 < R < 50 cm) of a new CMS tracker for Hl-LHC where neutrons will be the main dominated source of radiation damage. The estimated maximum total neutron fluence of 8×10^{14}/cm^2 at Hl-LHC for the central outer strip region of tracker (R ~ 40 cm) are reported [2]. The outer region is for strip detectors.

The radiation damage affects the Si sensors performance in terms of increases of leakage current due to increases of generation recombination centers, increases of full depletion voltage (V_{FD}) due to change of effective doping concentration, and degradation of charge collection efficiency (CCE) due to charge carries trapping [3]. The Double Junction (DJ) effects has been observed in irradiated Si sensors,

© Springer Nature Switzerland AG 2019
A. K. Srivastava, *Si Detectors and Characterization for HEP and Photon Science Experiment*, https://doi.org/10.1007/978-3-030-19531-1_3

which will give Double Peak (DP) electric field due to the occupation of deep traps. This phenomenon is also referred to type-inversion or space charge sign inversion (SCSI); Si sensors will undergo type inversion thus for higher CCE, sensors should be over depleted for the high electric field over the entire sensor volume.

In the frame of the CERN RD50 collaboration, a lot of progress has been made in recent years in order to improve the radiation hardness of Si sensors [4]. Defect engineering is one of R & D approach where several microscopic defects identified which has energy levels in the band gap of Si and have macroscopic effects on the Si sensor performance [5]. In mixed irradiations, the use of n-magnetic Czochralski (MCz) Si as a material prime candidate for the HL-LHC to improve the radiation hardness of Si sensors for the outer strip region of CMS tracker has been proposed [6]. Si sensor design for MCz material is still under serious investigations under RD50 collaboration and it has been already found that p-type Si sensors has improved CCE due to higher electron mobility of signal electrons than n-type and lack of type inversion effect after heavy irradiations.

In the framework of the CEC collaboration (within CMS), test structures of strixel geometry proposed and first attempt of simulation results on non-irradiated near-far strixel geometry have been reported [7]. The Si strixel sensor design with double metal layer with cross metal routing through via contact over the strip is also proposed as a new readout architecture for the large radii of a new CMS tracker for SHC whose performance has to verify using simulation approach.

3.2   Samples, Irradiation Campaign and Measurement Methodology

The MCz Si samples for n-type and p-type Si sensors are usually procured from Okmetic Ltd., Vantaa, Finland. These are 4 inches (100 mm) and 6 inches (150 mm) <100> Si wafer of different thickness 280, 290 and 300 microns. The resistivity of n-type is an order of 1×10^3 ohm-cm (> 500 ohm-cm) whereas for p-type is about 1.8 ohm-cm. The oxygen concentration [O] is almost same of 5–9×10^{17} cm^{-3} in both types of Si sensors.

A Different set of test structures of the MCz Si samples of n and p-type contains (CMOS capacitor, gated diodes, multi-guard diodes and microstrip/PAD detectors) were fabricated from Micron Semiconductors Ltd., ITC-IRST (Trento, Italy), IMP-CNM (Barcelona, Spain), Helsinki Institute of Physics (Helsinki, Finland) and BNL, USA. The test structures and Si detectors were characterized for Current-Voltage (I-V), Capacitance-Voltage (C-V) and Charge Collection Efficiency (CCE) measurements before and after irradiations at different temperatures.

For four types of beams (e, n, π, p) evolutions on Si sensors, sensors irradiation facilities are available. These are as follows: 24 GeV protons at CERN SPS; 23 GeV proton CERN PS; 200 MeV pions (π) at Paul Scherrer Institute (PSI), reactor neutrons at Josef Stefan Institute (JSI), Ljubljana Slovenia and at compact Cyclotron Forschungszentrum for 26 MeV proton in Karlsruhe (Germany).

The methodologies for experimental measurements (I-V, C-V and CCE) are as follows: firstly sensors will be irradiated at room temperature (RT = 20 °C) and then transported at low temperature of T < 5 °C to the measurement laboratory and the during storage T = −17 °C. The experiments are performed immediately after irradiations at room temperature of 20 °C (±2%) and after beneficial annealing of 60 min 80 °C. The CCE measurements are also performed after long term annealing at different low temperatures.

In the present review, samples (n/p Mcz Si) were taken;

Samples	Size	Thick
PAD diodes	0.25 cm^2	280, 290, 300 μm (4.9 × 10^{12}, 2.87 × 10^{12} cm^{-3})
	0.625 cm^2	
Microstip sensors	16 cm^2	300 ± 2 μm

For p-type Mcz Si sensor, the isolation among n$^+$ strips is obtained with two uniform p-spray (no p-stop used) implantation dose i.e. low (2 × 10^{12} cm^{-2}) and high (5 × 10^{12} cm^{-2}).

3.3 Device Simulation Approach and New Developments

The simulation of macroscopic radiation damage effects are carried out using a physical model of the Synopsys TCAD device simulator [8] which is based on Shockley-Read-Hall (SRH) recombination statistics that is applied to the multi-levels deep traps into the band gap of Si [9–11]. In my previous work, we have described n-MCz Si four level deep traps model (Table 3.1) for modelling of neutron radiation damage effects in Si sensors and it will fairly agree to experimental data [12]. For the n-MCz Si substrates, our model includes two acceptor levels located at $E_c − 0.46$ eV (E$_5$) and $E_v + 0.42$ eV (H152K) plus two donor level at $E_c − 0.1$ eV and $E_v + 0.36$ eV (C$_i$O$_i$). In order to get agreement with experimental data, I have increases the introduction rate of cluster related defect center E5 to 12.4 and finally tune the capture cross-section of deep traps. The main effects of E$_5$, H152K, E30K and C$_i$O$_i$ on macroscopic parameter's of the Si sensors are as follows: the increase of leakage current, negative space charges contribution, relevant for positive space charges and also itself for positive space charges.

The effective doping concentration (N$_{eff}$), leakage current at full depletion (V$_{FD}$), the occupancy of traps (n$_T$), and the emission coefficient for electrons (acceptors) e$_n$

Table 3.1 The four level n-MCz Si radiation damage model for neutron irradiation (as-irradiated)

Defect	Energy level [eV]	σ_n [cm^2]	σ_p [cm^2]	η [cm^{-1}]
E5	$E_C − 0.46$	3.0 × 10^{-15}	4.1 × 10^{-15}	12.4
H(152 K)	$E_V + 0.42$	3.05 × 10^{-13}	1.0 × 10^{-13}	0.06
C$_i$O$_i$	$E_V + 0.36$	1.64 × 10^{-14}	2.24 × 10^{-14}	1.10
E(30 K)	$E_C − 0.10$	2.77 × 10^{-15}	2.0 × 10^{-15}	0.017

and holes (donors) e_p can be calculated by following expressions (first order approximation) [13, 14];

$$\Delta N_{eff} = N_D + \sum \left(n_T(donors) \right) - \sum \left(n_T(acceptors) \right) \qquad (3.1)$$

$$I(V_{FD})(T) = qAW_N \left(e_p \Sigma n_T(donors) + e_n \Sigma n_T(acceptors) \right), \qquad (3.2)$$

where N_D is the doping concentration of the Si substrate, W_N is the depletion width, and T is the temperature.

$$n_T(T) = N_T \frac{e_{n,p}}{e_n + e_p} \qquad (3.3)$$

$$e_{n,p}(T) = c_{n,p}(T)N_{C,V}(T)\exp\left(\pm \frac{E_a(T) - E_{C,V}}{k_b T} \right) \qquad (3.4)$$

where $c_{n,p}$ ($c_{n,p} = v_{th}\sigma_{n,p}$) represents the capture coefficient of electrons and holes, respectively; $N_{C,V}$ represent the effective density of states of charge carriers in conduction band and valence band respectively; $E_{C,V}$ is the conduction and valence band energies, respectively; and E_a is the activation energy. The $\sigma_{n,p}$ is capture cross section of the electrons and holes.

3.4   Comparisons with Experimental Results

In this section, we have shown the comparisons of the simulated full depletion voltage and leakage current with the experimental results (see Fig. 3.1a and b) for neutron irradiated n-MCz Si sensor. It has been found that there is good agreement between simulations and experiment results for full depletion voltage (V_{FD}) and leakage current but theoretical calculations reproduce experimental V_{FD} but underestimate experimental current values due to first order approximation in Eq. (3.2). A further development in the aforesaid model/p-MCz is required for better understanding of radiation damage effects in the n/p MCz Si sensor.

3.5   Performance of MCz Si Sensors

There is a technological issue in both n-type and p-type MCz Si sensor designs: which strip isolation (p-stop/p-spray) in p-type, breakdown risk (near to cur edge, curvature of junction and edge of isolation) and optimal design of radiation hard Si sensors for SLHC. The performance characteristics of an irradiated detector are

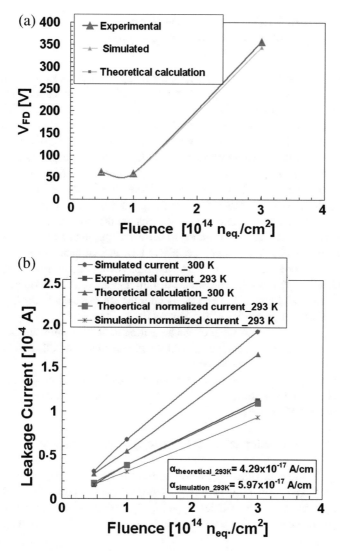

Fig. 3.1 (**a**) The full depletion voltage as a function of the fluence. (**b**) The leakage current as a function of the fluence [12, 15]

measured in terms of leakage current, full depletion voltage, and charge collection efficiency.

The performance of sensors should also be carefully checked for the interstrip capacitance which is a creation of crosstalk noise in the readout electronics.

3.6 Impact of Microscopic Defects on Macroscopic Parameters of Irradiated Si Sensors

When highly energetic ionising particles penetrating a sensor interact with the Si and generate electron–hole (e-h) pairs along their path that can be collected at the opposite polarity electrodes. If incident particles have enough energy ($E_k > 5$ eV), a lattice atom can be displaced from its original position (Primary Knock-on Atom or PKA) and two point defects are generated in the silicon lattice, namely an interstitial I atom and a vacancy V (referred to as a Frenkel pair). Because of their very high mobility, most of these pairs quickly recombine, but a significant proportion can interact with lattice impurities and migrate atoms and produce electrically active defects which have energy states in the band gap; after irradiation, shallow donor and acceptors will be introduced and thus it will change the "effective" doping concentration of the bulk Si,. This contributes to substrate type-inversion or space charge sign inversion (SCSI). These electrical active defects can behave like Generation-Recombination (G-R) centres, or acting as traps for charge carriers (reduction in charge collection efficiency). The effects of bulk radiation damage on macroscopic parameters are well known [11] and the deep traps are responsible for the degradation of the electrical properties of the Si sensors.

For annealing of Si sensors, beneficial annealing of 60 °C 80 min annealing (short time) and then reverse annealing (long time) will take place and this will change the macroscopic properties of the Si sensors.

The Hamburg model [1] is used to calculate the effective doping concentration immediately after irradiation, and after annealing (time, temperature) in neutron, proton irradiated Si sensors.

$$\text{The Hamburg model is } \Delta N_{eff}(\Phi, t) = N_{eff,0} - N_{eff}(\Phi, t) \tag{3.5}$$

In Eq. 3.5, the following parameters are used: t = annealing time, $\Phi = \Phi_{eq}$ (1 MeV neutron equivalent fluence) $N_{eff,0}$ = doping concentration ($N_D - N_A$) before irradiation, in normal n-type diodes $N_A \ll N_D$, $N_{eff}(\Phi, t)$ = "doping concentration" (effective space charge conc.) after irradiation with fluence Φ at annealing time t.

$$\Delta \cdot N_{eff}\left(\Phi, t \, (80 \deg C)\right)$$
$$= -N_{a0} \exp(-t/\tau_a) + N_{eff0} \exp(-c\Phi) -$$
$$- \Sigma g_{CA} \cdot \Phi + \Sigma g_{CD} \cdot \Phi + N_{Y1}(1 - \exp(-t/\tau_{Y1})) + N_{Y2}(1 - 1/(1 + t/\tau_{Y2})), \tag{3.6}$$

where $N_{eff0} \exp(-c\Phi)$: donor removal, for large Φ complete removal.

If all Phosphorus-donors combine with vacancies V during the process then this part should be equal to the E-center formation

$-g_C \cdot \Phi$ \qquad acceptor generation (minus sign! hence acceptors)

$+g_{CD} \cdot \Phi$ \qquad donor generation (plus sign! hence donors)

$-N_{a0}\exp(-t/\tau_a)$ component of those acceptors, initially irradiated but annealing out

$-N_Y(t)$ reverse annealing (generation of acceptors during annealing).

The understanding of the physical structure of these defects and their effect on the macroscopic sensor properties for the possible improvements of the sensor performance through material/defect engineering is under way in WODEAN collaboration.

The identification of microscopic defects in p-MCz Si is still under way in the framework of the CERN RD50 collaboration and WODEAN collaboration, and here we are summarizing the known defects and its impact on macroscopic sensor parameters.

n-MCz Si Sensors

Figure 3.2 shows the TSC spectra after forward injection in 1 MeV neutron irradiated n-MCz Si diode, $\Phi = 5 \times 10^{13}$ cm^{-2} [13]. In this fig, three hole deep traps labelled H (116 K), H (140 K), and $H \lesssim 152$ K) proved to have an electric-field-enhanced emission characteristic of Coulombic wells. This is finally pointing out that the hole H (116 K), H (140 K), and $H \lesssim 152$ K) are acceptors in the lower part of the silicon band gap. They contribute with their full concentration as negative space charge to Neff 0 in n-type silicon diodes and shows reverse annealing. These levels do not form with γ radiation and are therefore cluster defects. It has been found that the concentration of these deep level defects increases with long time annealing corresponding with negative space charge build-up (N_{eff} change).

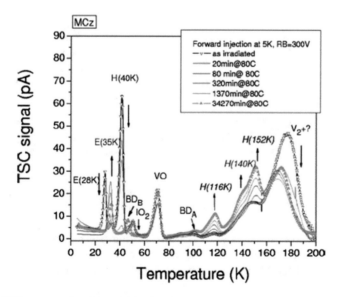

Fig. 3.2 TSC spectra after forward injection in 1 MeV neutron irradiated n-MCz Si diode, $\Phi = 5 \times 10^{13}$ cm^{-2} [13]

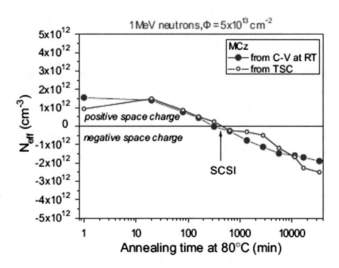

Fig. 3.3 For each annealing steps, the depletion voltage was determined by means of a C-V characteristics and a TSC scan of an irradiated n-MCz Si diode, $\Phi = 5 \times 10^{13}$ cm^{-2}. The change matches the rise of the defect levels H (116 K), H(140 K), H(152 K) [13]

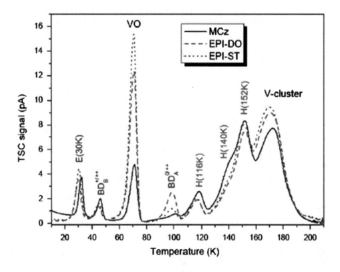

Fig. 3.4 Comparison of TSC spectra measured on MCz, EPI-DO and EPI-STdiodes after neutron irradiation with $\Phi = 5 \times 10^{13}$ cm^{-2} after isothermal annealing for 16,980 min at 80 °C [13]

Figure 3.3 shows the change in effective doping concentration with annealing time at 80 °C. At long annealing time, the concentrations of these defect levels H (116 K), H (140 K), H (152 K) increases (see in [13]) with long time annealing corresponding with negative space charge build-up (N_{eff} also increases).

A comparison of TSC spectra obtained in EPI-ST, EPI-DO and MCz diodes after the same neutron irradiation of 5×10^{13} cm^{-2} is displayed in Fig. 3.4 as example for

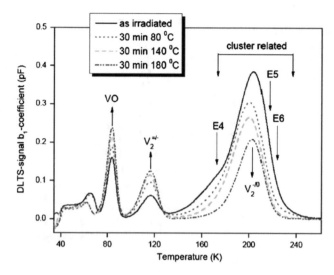

Fig. 3.5 DLTS spectra measured on a n-MCz diode after neutron irradiation with $\Phi = 3 \times 10^{11}$ cm^2, examples of isochronal annealing steps between 20 and 200 °C [13]

long-term annealing at 80 °C. The shallow donor positive charge E (30 K) defect is clearly identified in Fig. 3.4, and acting as a trap for electrons. Also this defect was only detected after hadron irradiation (and to some extent after energetic electron irradiation), but not after γ-irradiation. The E (30 K) is responsible for positive space charge [13] and therefore partly compensate the negative space charge induced by the acceptors. In contrast to deep acceptors, the introduction rate of E30 K is higher in proton irradiated sensors than neutron irradiations. For proton irradiation the effective space charge remains positive while after neutron damage type inversion (negative space charge) is observed.

The DLTS signal shows the evolution of E4, E5 and E6 (cluster related defect centre) after annealing. It is responsible for the increase of leakage current in irradiated n-MCz Si sensor (see Fig. 3.5).

The radiation hardness of Si sensors can be improved in the mixed irradiation case of n-MCz Si by compensation of the cluster-related deep acceptors (neutron irradiations) with enhanced donor generation (proton irradiations).

The physical structures of the microscopic defects are unknown for the deep acceptors and shallow donors. A lot of R & D work is on going under WODEAN and CERN RD50 collaboration for the development of the radiation hard Si detectors of the HL-LHC.

The n in p MCz Si sensors may be interesting for HL-LHC for the outer region of the phase 2 upgrade of the CMS tracker detector. There is no significant microscopic defect known to find the correlation between microscopic and macroscopic parameters.

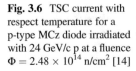

Fig. 3.6 TSC current with respect temperature for a p-type MCz diode irradiated with 24 GeV/c p at a fluence $\Phi = 2.48 \times 10^{14}$ n/cm^2 [14]

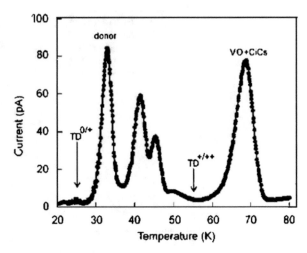

In Fig. 3.6, there is donor peak at 30 K but the effect on macroscopic parameters is not yet confirmed in p-MCz Si sensor. A lot of work is needed for p-MCz Si for identification of microscopic defects [14].

3.6.1 I/V, C/V Behaviour in Irradiated Si Sensors

The leakage current and full depletion voltage is measured in the different irradiated Si sensor (Fz, n/p MCz, Epi-n) using I/V, C/V experimental set up. In Fig. 3.7, the leakage current per unit volume as a function of the fluence for (Fz, n/p MCz, Epi-n) sensor irradiated with reactor neutrons and protons are shown. It has been found that the leakage current is increases linearly with the fluence as expected for all sensors and that is independent of the materials, type of sensor, and also type of irradiations. Therefore we can say that the current related damage constant (α) is same for both type (n/p MCz) and it is order of 3.7–4.4×10^{-17} A/cm after beneficial annealing.

It is expected that α should be also same for mixed irradiation (n + p) samples [6]. The annealing of leakage current is not shown here.

The effective doping concentration versus fluence is shown in Fig. 3.8 for n-MCz Si for n and p irradiations (around 8.6×10^{14} n/cm^2) but not for p-MCz Si.

It has been found that the depletion voltage (effective doping concentration) is higher in neutron irradiated Si sensor ($>1 \times 10^{14}$ n/cm^2) than proton irradiated.

The depletion voltage of neutron irradiated sensors as function of annealing time is shown in Fig. 3.9 (Smart Italian project). It is observed that after annealing of 80 °C up to 600 min, the reveres annealing growth is higher in p-MCz Si sensors than n-type and depletion voltage is higher than n-type MCz Si sensor. It has been also found that the full depletion voltages are almost constant after 200 min

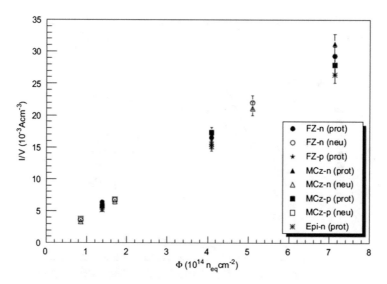

Fig. 3.7 Reverse current per unit volume as a function of the n_{eq} irradiation fluence for 26 MeV proton and reactor neutron irradiated FZ, MCz and EPI devices. All measurements have been carried out at 20 °C, at $V_{bias} = V_{FD} + 50$ V, after an annealing treatment of 8 min at 80 °C [6]

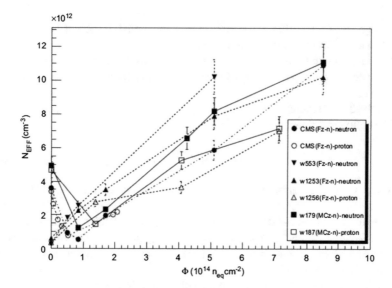

Fig. 3.8 Effective doping concentration N_{eff} vs. fluence Φ for n-type FZ and n-MCz diodes irradiated with 26 MeV protons and with reactor neutrons [6]

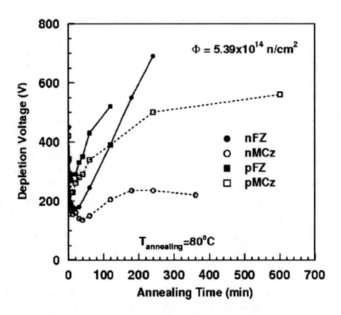

Fig. 3.9 V_{depl} evolution as a function of annealing time

annealing for n, p- MCz Si sensor. It is known that reverse annealing is the significance of less radiation hard Si sensors.

The full depletion voltage (V_{fd}) versus $\Phi_{eq.}$ Is shown in Fig. 3.10 a and (b) for MCz-n and MCz-p type Si samples after mixed irradiation after 60 °C and 80 min annealing.

In mixed irradiations,

The equivalent fluence is given by

$$\Phi_{eq.} = \kappa_p \varphi_p + \phi_{eq.}, \tag{3.7}$$

where κp is the hardness factor for proton irradiations and Φ_p is the proton fluence.

In mixed irradiated n-MCz Si samples (firstly proton irradiated with 1.85×10^{14} p/cm^2 + 2×10^{14} n/cm^2 reactor neutrons), the V_{fd} decreases due to compensation effects [6] whereas in p-MCz Si (mixed); the V_{fd} increases, the reason is the higher introduction rate of E30 K.

The V_{fd} evolution as a function of annealing time is shown in Fig. 3.11.

In n-MCz Si sensor (mixed irradiation), the V_{fd} initially decreases up to 20 min annealing and constant after 20 min annealing. Whereas in p-MCz Si sensor, the V_{fd} initially decreases up to 100 min annealing and it is clear from Fig. 3.11 that the reverse annealing growth strongly increases after 500 min annealing. It should be noted that for approx same composition of mixed irradiations, the lower V_{fd} is obtained in n-type MCz Si than p-type MCz Si sensor.

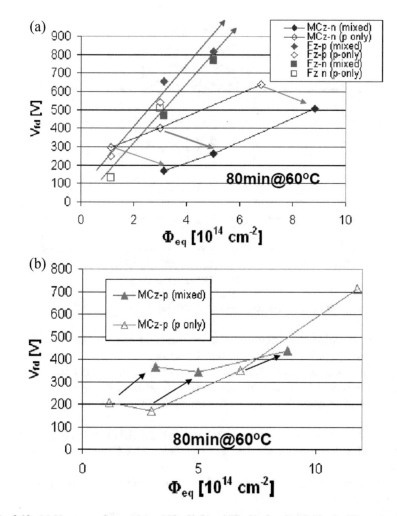

Fig. 3.10 (a) V_{fd} versus fluence (a) n-MCz Si (b) p-MCz Si after 60 °C 80 min [6]

Therefore, it is recommended to check the performance of p-MCz Si sensors for the different compositions of mixed irradiation for the HL-LHC.

3.6.2   CCE in Irradiated Si Sensors

The most probable collected charge (CCE = collected charge/charge produced) versus bias voltage is shown in Fig. 3.12 for MCz-n (mixed) and MCz p (mixed) Si sensor. It is observed that CCE in n-MCz is better than p-MCz Si sensor. It is

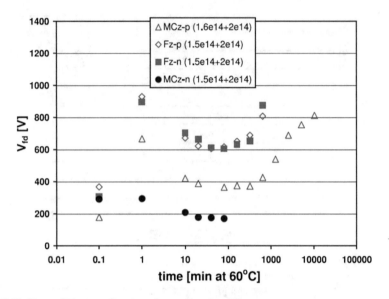

Fig. 3.11 V_{fd} evolution as a function of annealing time [6]

known that trapping is dominant factor for reduction of CCE. In Fig. 3.12, CCE reduces with increasing fluence of proton + neutrons.

The CCE shows annealing behavior for MCz-n Si (mixed). It is observed that the CCE improves after long term annealing at 60 °C (Fig. 3.13).

It is should be noted that CCE do not anneal (see Fig. 3.14) in MCz-p Si sensor (mixed).

The CCE for n in p MCz sensors are shown in Fig. 3.15 for high neutron fluences $(5 \times 10^{14}$–2.2×10^{16} $n_{eq}./cm^2)$. It is observed that the CCE reduces with increasing fluence.

In Fig. 3.16, CCE also do not anneal for proton and also neutron irradiated Si sensor n in p MCz Si sensors [17].

This is to note down that there is no evidence reported for avalanche multiplication effect (CCE > 1) in irradiated n/p MCz Si sensor.

3.6.3   Interstrip Capacitance in Irradiated Si Sensors and Test Beam Results for S/N

The interstrip capacitance (C_{it}) is measured by the SMART collaboration [18, 19] to evaluate the expected cross-talk noise in the readout electronics. Before irradiation of the sensor, C_{it} reaches to a minimum value between 0.5 and 1.2 pFcm⁻¹, depending on the sensor geometry, once the sensors are over-depleted. The n-type MCz and the FZ microstrip detectors have comparable C_{it} after irradiation. In n-type MCz Si sensors, C_{it} did not change significantly with the fluence. On the contrary, in

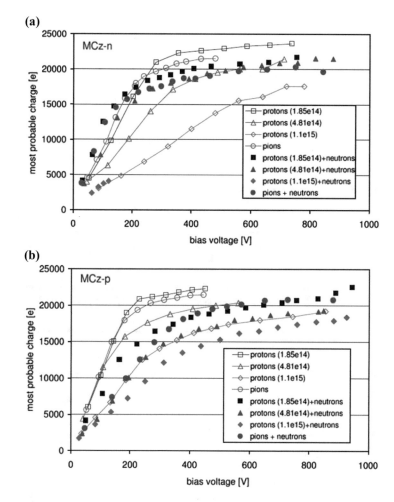

Fig. 3.12 (**a**) n-MCz Si and (**b**) p-MCz Si. Dependence of collected charge on voltage for mixed irradiated and control diodes. The fluence for proton irradiated detectors given in brackets [6]

p-type sensors C_{it} decreased with the fluence and reached the n-type values in sensors irradiated to 4×10^{14} n_{eq}/cm^2.

The more R & D work is required for the C_{int} analysis in neutron, proton, and mixed irradiated n, p-MCz Si sensors.

The Test beam results of n-MCz Si strip sensor of an area of 16 cm^2 are described here in brief.

The Si strip sensors of area 4 cm^2 × 300 ± 2 μm and a nominal resistivity of 900 Ωcm are fabricated at Helsinki University of Technology, Micronova Centre for Micro and Nanotechnology. The detector has 768 channels, and the detector design strip pitch was 50 μm, strip width 10 μm, and the strip length 3.9 cm. The detectors were irradiated with 25 MeV proton at Karlsruhe (Germany) and 3–45 MeV

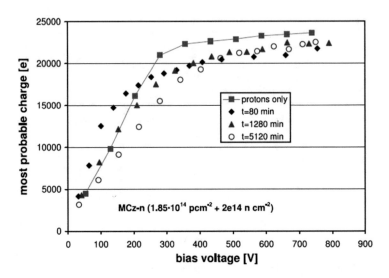

Fig. 3.13 Dependence of charge collection efficiency on voltage during long term annealing at 60 °C for mixed (23 GeV protons and reactor neutron) irradiated MCz–p pad detectors [6]

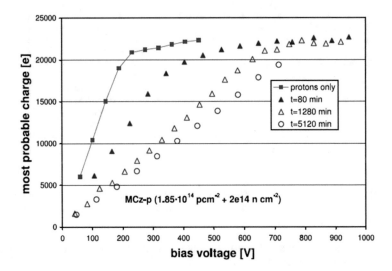

Fig. 3.14 Dependence of charge collection efficiency on voltage during long term annealing at 60 °C for mixed (23 GeV protons and reactor neutron) irradiated MCz–p pad detectors [6]

neutrons at UCL, Louvain up to a fluence of 3×10^{15} $n_{eq}./cm^2$ either with proton or mixed p + n (different composition).

The test beam was carried out the using CERN H2 beam line using a SiBT (Silicon Beam Telescope) that determine (Signal/Noise = S/N > 10 for the n-MCz Si strip sensors for the fluence of 1×10^{15} $n_{eq}./cm^2$.

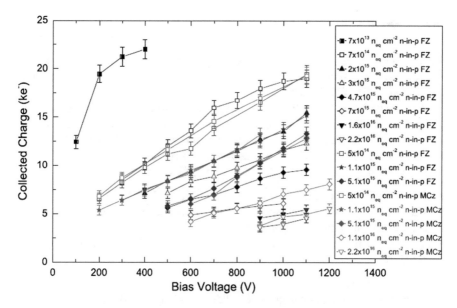

Fig. 3.15 Collected charge as a function of the bias voltage for the different sensors [16]

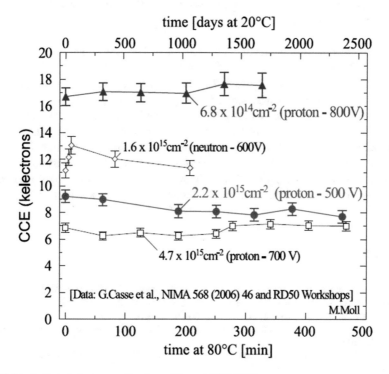

Fig. 3.16 Collected charge as a function of time at 80 °C [17]

3.7 Conclusion and Outlook

In this paper, we reviews of all published important works are summarized for the understanding of the important candidate (n/p MCz Si) for the outer tracker region of the upgrade of the CMS tracker for Hl-LHC.

It has been found that p-type Si has significantly improved performance for all prospects than n-type in the single irradiation environment. But, for mixed irradiations, n-MCz is better than p-MCz Si sensor so it should be again carefully checked for different composition of charged hadrons and neutrons.

The following things are concluded for n in p Si sensors:

- Trapping should be reduced or control for sufficient high CCE.
- The knowledge of physical structure of cluster defects is necessary for improvement in radiation hardening of devices.
- n in p MCz Si could be choice for the inner region of HL-LHC and it should be noted that

  – no type inversion
  – design risk (optimal design for breakdown voltage >600 V)
  – CCE do not anneal
  – high bias voltage (may be breakdown will occur). This is because of the high reverse annealing growth.

The increased instantaneous and integrated luminosity anticipated for the HL-LHC (2026) is reported in [20, 21]. The inner tracking detector will be upgraded for HL-LHC radiation conditions [22]. The silicon strip sensors in the upgraded trackers will consist of n-type strips in p-type substrates (n-in-p) and replace the previously used (p-in-n) technology. This is driven by radiation hardness considerations, as it was demonstrated that ionizing and displacement damage have a less detrimental impact on detector performance for n-in-p devices than for p-in-n devices. For the pixel detector layers various sensor options are feasible with thin planar n-in-p sensors bump bonded to the readout ASIC (Hybrid Pixel Detectors) being the baseline solution, replacing the previously used n-in-n sensor technology.

References

1. Fretwurst, E.: Recent advancements in the development of radiation hard semiconductor detectors for S-LHC. Nucl. Instr. Methods Phys. Res. A. **A552**, 7–19 (2005)
2. Huntinen, M.: SLHC Electronics workshop (2004), CERN
3. Lindstrom, G.: Radiation damage in silicon detectors. Nucl. Instr. Methods Phys. Res. A. **A512**, 30–43 (2003)
4. RD50 Status Reports CERN-LHCC-2003-2004-2005-2007 [Online]. Available http://rd50.web.cern.ch/rd50/
5. Moll, M.: Ph.D. thesis. Radiation Damage in Silicon Particle Detectors, University of Hamburg, Germany, 1999, DESY-THESIS-1999-40, ISSN 1435-8085

6. Kramberger, G., Cindro, V., Dolenc, I., Mandic, I., Mikuz, M., Zavrtanik, M.: Performance of silicon pad detectors after mixed irradiations with neutrons and fast charged hadrons. Nucl. Instr. Methods Phys. Res. A. **609**, 142–148 (2009)
7. Militaru, O. (CEC collaboration): Simulation of electrical paramaters of new design of SLHC silicon sensors for large radii. Nucl. Instr. Methods Phys. Res. A **617**, 563–564 (2010)
8. Moscatelli, F., Santocchia, A., Passeri, D., Bilei, G.M., MacEvoy, B.C., Hall, G., Placidi, P.: An enhanced approach to numerical modelling of heavily irradiated silicon devices. Nucl. Instr. Methods Phys Res. B. **186**, 171–175 (2002)
9. Passeri, D., Baroncini, M., Ciampolini, P., Bilei, G.M., Santocchia, A., Checchucci, B., Fiandrini, E.: TCAD-based analysis of radiation hardness in silicon detectors. IEEE Trans. Nucl. Sci. **45**(3), (1998)
10. Passeri, D., Ciampolini, P., Bilei, G., Moscatelli, F.: Comprehensive modeling of bulk-damage effects in silicon radiation detectors. IEEE Trans. Nucl. Sci. **48**, 1688–1693 (2001)
11. Moscatelli, F., Santocchia, A., Passeri, D., Bilei, G.M., MacEvoy, B.C., Hall, G., Placidi, P.: An enhanced approach to numerical modelling of heavily irradiated silicon devices. Nucl. Instr. Methods Phys Res. B. **186**, 171–175 (2002)
12. Srivastava, A.K., Eckstein, D., Fretwurst, E., Klanner, R., Steinbrück, G.: Numerical modelling of the Si sensors for the HEP experiments and XFEL. Presented at RD09 conference on 30 September – 02 October 2009
13. Pintilie, I., Fretwurst, E., Lindström, G.: Cluster related hole trap with enhanced-field emission – the source for long term annealing in hadron irradiated Si diodes. Appl. Phys. Lett. **92**, 024101 (2008)
14. Lang, D.V.: Bräunlich, P. (ed.): Thermally Stimulated Relaxation in Solids, vol. 179, pp. 3–128. Springer, Berlin
15. Synopsys, Inc., TCAD software. URL http: www.synopsys.com/products/tcad/tca.html
16. Affolder, A., et al.: Silicon detectors for the sLHC. Nucl. Instr. Methods Phys. Res. A, 658(1) (1 December 2011)
17. Casse, G.: Overview of the recent activities of the RD50 collaboration on radiation hardening of semiconductor detectors for the sLHC. NIMA. **598**, 54–60 (2009)
18. Chatterji, S., Bhardwaj, A., Ranjan, K., Namrata, Srivastava, A.K., Shivpur, R.K.: Analysis of interstrip capacitance of Si microstrip detector using simulation approach. Solid State Electron. **47**(9), 1491 (2003)
19. CERN. Website http://www.cern.ch/rd50/
20. Apollinari, G., et al.: High-Luminosity Large Hadron Collider (HLLHC), technical design report v. 0.1. CERN, Tech. Rep. CERN-2017-007-M (2017)
21. The High Luminosity LHC (HL-LHC) project. [Online] Available http://hilumilhc.web.cern.ch/about/hl-lhc-project
22. Moll, M.: Displacement damage in silicon detectors for high energy physics. IEEE Trans. Nucl. Sci. **65**(8), 1561–1582 (2018). https://doi.org/10.1109/TNS.2018.2819506

Chapter 4
Development of Radiation Hard Pixel Detectors for the European XFEL

The Linac Coherent Light Source (LCLS) [1] at the SLAC National Accelerator Laboratory, U.S.A. commissioned and operated in 2009. The great Prof. (Dr.) John Madey [2] had invented the Free-Electron Laser at Stanford University, U.S.A. and after 30 years of its invention DESY, Hamburg, Germany has planned to set up a new fourth-generation of hard X-ray sources FEL Experiment for the wide range of challenging applications in the world. The X-ray Free-Electron Lasers (XFEL) provide femtosecond-duration and a high degree of spatial coherence pulses of hard X-rays with a peak brightness approximately one billion times greater than is available at synchrotron radiation facilities (see Fig. 4.1). With the FEL light, the functional material at an interatomic distance and time scales of an atomic motion [4] can be explored. In the material science, as an example, inducing transient structures using ultrafast low wavelength light pulses that can considerably change material properties.

By using X-ray pulses of a FEL the imaging of processes will be possible in these systems. In the biological sciences, time-resolved X-ray crystallography will experience a dramatic increase using FEL sources. The extreme instantaneous brightness of the pulses will allow to shrink the crystal sizes all the way down possibly to single molecules, giving three-dimensional movies of conformational dynamics and chemical reactions, and allowing the imaging of macromolecules that cannot be crystallized.

4.1 Introduction

The European X-ray Free-Electron Laser (XFEL) [5], which is currently under construction in Hamburg, Germany, and which will enter user operation in 2016, will go one step further compared to LCLS or SACLA (SPring-8 Ångstrom Compact free electron laser) [6] by using superconducting RF cavities. This will result in an unique bunch time pattern in contrast to the 120 evenly spaced pulses produced at the LCLS. The XFEL pulses will be delivered in trains (see Fig. 4.2) of typically

© Springer Nature Switzerland AG 2019
A. K. Srivastava, *Si Detectors and Characterization for HEP and Photon Science Experiment*, https://doi.org/10.1007/978-3-030-19531-1_4

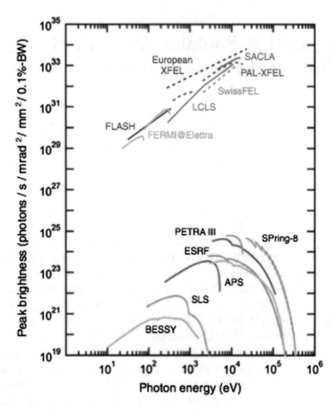

Fig. 4.1 Peak brightness (photons/(s.mm^2.mrad^2.0.1%BW)) as function of photon energy for planned and existing FELs compared to synchrotron sources [3]

2700 pulses with more then 1012 photons/pulse at 4.5 MHz followed by a gap of 99.4 ms. The high instantaneous intensity, short pulse duration and high repetition rate will pose very demanding requirements for imaging detectors. The European XFEL initiated therefore three independent detector development projects to meet the requirements for the different instruments which will be installed at initially 6 beam lines.

The Adaptive Gain Integrating Pixel Detector (AGIPD) [7, 8] system, which is one of these XFEL projects, was installed and commissioned in August 2017 for the use in the SPB (Diffraction of Single Particles and Biomolecules [9]) and MID (Materials Imaging and Dynamics [10]) instruments. It is a hybrid-pixel detector (HPAD) system with 1024 × 1024 p$^+$-pixels of dimensions (200 μm)^2, built of 16 p$^+$-n silicon sensors, each with a sensitive area of 10.52 cm × 2.56 cm and a thickness of 500 μm. The particular requirements (see Fig. 4.3) are a dynamic range of 0, 1 to more than 10^4 photons of 12.4 keV per pixel for a pulse duration of less than 100 fs, negligible pile-up at the XFEL repetition rate of 4.5 MHz, and operation for X-ray doses up to 1 GGy in 3 years [12]. In addition, the sensors should have a good detection efficiency for X-rays with energies between 5 and 20 keV, and minimal inactive regions at their edges.

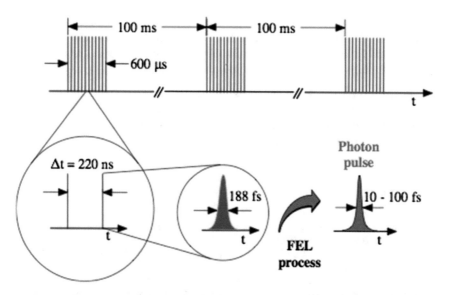

Fig. 4.2 European XFEL bunch time pattern

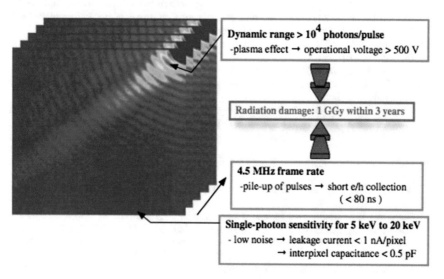

Fig. 4.3 Challenges of silicon detectors used at the European XFEL for imaging experiments. Diffraction pattern taken from [11]

The implication of the high number of photons per pixels, which cause the so-called e-h plasma effect, was studied by J. Becker [13] with the main result for the AGIPD that an applied voltage of well above 500 V is required. The surface radiation damage were studied by A.K. Srivastava [14] and J. Zhang [15] on test structures to extract surface radiation damage parameters and the microscopic parameters that were used by A.K. Srivastava [14] in the TCAD simulation to get the first optimal design of n$^+$n pixel detectors. T. Poehlsen [16] has studies the charge losses in the Si detector close to the Si-SiO$_2$ interface. The impact of the surface radiation damage on the detector can be summarize in the following way:

1. Decrease of the avalanche breakdown voltage.
2. Increase of the surface leakage current at an Si-SiO$_2$ interface.
3. Increase of the full depletion voltage and interpixel capacitance.
4. Occurrence of charge losses close to the Si-SiO$_2$ interface.

References

1. Emma, P., et al.: First lasing and operation of an Ångstrom-wavelength free-electron laser. Nat. Photonics. **4**(9), 641–647 (2010)
2. Madey, J.M.J.: Stimulated Emission of Bremsstrahlung in a periodic magnetic field. J. Appl. Phys. **42**(5), 1906 (1971)
3. Schmüser, P., Dohlus, M., Rossbach, J., Behrens, C.: Free-Electron Lasers in the Ultraviolet and X-Ray Regime: Physical Principles, Experimental Results, Technical Realization. Springer, Berlin (2014)
4. Chapman, H.N.: Coherent imaging with x-ray free-electron lasers. In: Angst, M., Brückel, T., Richter, D., Zorn, R. (eds.). Scattering Methods for Condensed Matter Research: Towards Novel Applications at Future Sources. Forschungszentrum Jülich (2012)
5. Altarelli, M., et al.: The European X-Ray Free-Electron Laser, Technical design report. DESY, Hamburg (2006)
6. Sacla. URL: http://xfel.riken.jp/eng/index.html
7. AGIPD. URL.: http://photon-science.desy.de/research/technical_groups/detectors/projects/agipd/index_eng.html
8. Henrich, B., et al.: The adaptive gain integrating pixel detector AGIPD a detector for the European XFEL. Nucl. Instr. Methods A. **633**(Supplement 1), S11–S14 (2011)
9. Mancuso, A.: Conceptual Design Report Scientific Instrument SPB, Technical Report TR-2011-007 (2012)
10. Madsen, A.: Conceptual Design Report Scientific Instrument MID, Technical Report TR-2011-008 (2012)
11. Seibert, M.M., et al.: Single mimivirus particles intercepted and imaged with an X-ray laser. Nature. **469**(7332), 78–81 (2012)
12. Graafsma, H.: Requirements for and development of 2 dimensional X-ray detectors for the european X-ray Free Electron Laser in Hamburg. J. Instrum. **4**(12), P12011 (2009)
13. Becker, J.: Signal development in silicon sensors used for radiation detection. Ph.D. thesis, Universität Hamburg, Hamburg (2010). DESY-THESIS-2010-033
14. Srivastava, A.K., et al.: Numerical modelling of Si sensors for HEP experiments and XFEL (POS RD09) 19 (2010)
15. Zhang, J.: X-ray Radiation Damage Studies and Design of a Silicon Pixel Sensor for Science at the XFEL. Ph.D. thesis, Universität Hamburg, Hamburg (2013). DESY-THESIS-2013-018
16. Poehlsen, T.: Charge Losses in Silicon Sensors and Electric-Field Studies at the Si–SiO$_2$ Interface. Ph.D. thesis, Universität Hamburg, Hamburg (2013). DESY-THESIS-2013-025

Chapter 5
T-CAD Simulation for the Designing of Detectors

5.1 What Is Computer Simulation Approach?

A simulation is the execution of a model, represented by a computer program that gives information about the system being investigated. The computer simulation approach of analyzing a model is opposed to the analytical approach, where the method of analyzing the system is purely theoretical. Simulation approach is more reliable and provides more flexibility and convenience.

5.1.1 Why Simulation?

The reduction in active device dimensions to micron and submicron geometries has resulted in an intimate coupling of the process conditions and device behaviour to a degree unknown a few years ago. It becomes more and more difficult to develop new processes due to the inherent complexity of semiconductor device fabrication. The use of Computer-Aided Design (CAD) tools has emerged as a very elegant mechanism to aid process and device engineer in their task of finding an optimum process and hence proven to be invaluable in the development of new technologies.

Traditionally, a new technology development is guided by an experimental "trial-and-error" approach. Starting with an existing process, certain steps in the process are changed, together with the structural dimensions. The modified process is then used to fabricate several lots of a device. However, this approach requires many iterations to optimize a new process, and fabricating one lot in a modern process can cost considerable amount of money and consume weeks or even months of effort. The use of accurate simulation tools in the proper computing environment, on the other hand, allows for comparatively inexpensive and time-saving "computer experiment". Reduction of both optimization time and prototypization expenses is therefore expected from the adoption of device simulation.

© Springer Nature Switzerland AG 2019
A. K. Srivastava, *Si Detectors and Characterization for HEP and Photon Science Experiment*, https://doi.org/10.1007/978-3-030-19531-1_5

5.1.1.1   T-CAD Tools: Process and Device Simulation

The goal of computer simulation tools is to predict the outcome of process steps, and thereby reduce the guesswork in developing a process sequence. T-CAD tool is the numeric simulation of semiconductor processes and devices. T-CAD tool is the utilization of computational methods, integrated into software tools, for the design and analysis of semiconductor devices and their fabrication processes.

In simulations, the device is modeled through a binary representation of its properties, for example, its geometry, material, dopant concentration, and temperature distribution in two or three dimensions. The complete process flow during manufacturing of the real device is represented by a series of numerical simulation tools, i.e., by the process simulators and the device simulators. Process simulation deals with all aspects of device fabrication. With the proper input parameters (processing steps, layout geometry) process simulation determines the details of the resultant device structure, including the boundaries of the different materials of the structure and the distribution of impurity-ions within the structure.

The output of the process simulation, together with the applied bias voltage is the input to a device simulation program. Important insight can be gained by analyzing the behavior of the internal variables such as electrostatic potential, electric field, carrier densities etc.

Finally, the semiconductor equations are solved in the generated device by a device simulator which computes the electrical characteristics of the device like breakdown voltage, leakage current etc.

History of T-CAD   The history of commercial T-CAD tools began with the formation of Technology Modeling Associates (TMA) in 1979. The software were an outgrowth of research done at Stanford University, California under Dutton and Plummer. The most famous of the Stanford T-CAD software programs are SUPREM and PISCES. SUPREM-3 is one-dimensional whereas SUPREM-4 is two dimensional process simulators. PISCES is the corresponding two-dimensional device simulator.

In the present work, simulations are carried out using two commercial software packages namely–TMA-SUPREM-4 version 1999.4 [1] for process simulation and TMA MEDICI version 2000.4 [2] for device simulation (both two-dimensional). With the help of these packages, we have been able to optimize various geometrical and physical parameters to improve the device characteristics of Si microstrip detectors.

5.2   TMA SUPREM-4: The Process Simulation Program

Salient features of TMA SUPREM-4 version 1999.4 are as follows:

  (i) It is 2-D software program for simulating the process steps involved in the fabrication of the devices.
 (ii) It incorporates diffusion rates for various impurities.
(iii) The information provided by the process simulator includes:

   – Diffusion profile of the dopants.
   – Distribution of impurities etc.

 (iv) Point defects and extended defects can be incorporated.
  (v) It calculates the electrical characteristics such as sheet resistance etc.

Description TMA SUPREM-4 uses a simulation grid which can be refined. We can adjust the initial specifications of the wafer like resistivity, length, width of the device and also incorporate the process flow along with suitable physical models. The output provides the value of various process parameters, for example thickness of oxide after oxidation, sheet resistance, junction depth, junction capacitance and the exact doping profile etc.

Physical Models The physical models used in the process simulator are the solid solubility, dopant clustering, dopant-defect clustering, small dopant-defect clustering model, and generalized proposed model for the formation and dissolution of {113} defects [3] etc. In addition, the model of Huang and Dutton for the interaction between interstitials and dislocation loops is also available [4].

5.3   TMA MEDICI: The Device Simulation Package

Salient features of TMA-MEDICI, version 2000.4 are as follows:

 • It simulates the electrical characteristics of semiconductor devices like leakage current, breakdown voltage etc.
 • It solves Poisson's equation, the electron and hole current continuity equations, energy balance equations and the lattice heat equations for holes and electrons.
 • It includes various device related models [5–8].
 • It includes generated carriers due to impact ionization.
 • The potential contours, impact ionization generation rate, electric field lines and current flow lines can be plotted, so that the exact location where the breakdown occurs can be identified.

Description TMA MEDICI needs a defined set of node points (maximum 20,000) or mesh to perform a numerical calculation. Then, the models and solution algorithms have to be specified. This also includes specifying parameters and selecting features. Generally, the command input file needs to have a specific structure/organization, which is ordered in four groups as follows:

- Mesh structure specification
- Coefficients and Material parameters specification
- Output solution specification
- I/O statements

An example of a typical input MEDICI program is shown in Fig. 5.1.

Physical Models In simulation, it is crucial to choose the correct boundary conditions for realizing practical realistic conditions. Dirichlet boundary conditions are imposed on the Ohmic contacts, whereas homogeneous (reflecting) Neumann conditions are implemented on the non-contacted edges of the structure. To accurately predict junction breakdown voltage, junction curvature effect causing higher electric field at the device corner is included in the program. It includes various types of device related models as mentioned above [5–8].

5.4   Validation of Process and Device Simulation Packages

The process and device simulators are required to be rigorously calibrated against experimental data to validate the simulation results.

In order to validate the accuracy of the process/device simulation programs used in the present work, experimental data available in the literature [9–12] is simulated and the results are compared. It is found that the difference between the observed and simulated values of process and device parameters is very small, thus validating the accuracy of the simulation packages. A more advanced simulation program using Synopsys TCAD program for current-voltage and capacitance-voltage is shown in Fig. 5.2a and b.

```
$ An example of a typical MEDICI program
$ Create an initial Simulation Mesh and save the mesh file
MESH   OUT.F=<filename>
X.MESH  X.MAX=   H1=  H2=  H3=
Y.MESH  N=  location=   XY ratio=
Y.MESH  Y.MAX=   H1=   H2=   H3=
$
$ Region Definition
REGION   NAME= <name>
$
$ Electrode Definition
ELECTR   NAME=<name>   x.min= x.max= y.min= y.max=
$
$ Specify dopant Impurity Profiles
PROFILE  N-TYE   N.PEAK= <value>   x.min= x.max= y.min= y.max=
$
$ Specify fixed oxide charge at the Si-SiO2 Interface (say 2x10^11 cm^-2)
INTERFAC   Qf=2E11
$
$ Material Specifications, All parameters are set by default values, but I want to show
$ how to specify user defined values. If I want to change the band-gap of Silicon from
1.08eV
$ (default) to another value, say 1.12eV, we may write
MATERIAL  REGION=<name> EG300 =1.12
$ Define Models, example Shockley Read Hall recombination with concentration
$ dependent lifetimes, concentration dependent mobility and impact ionization model
MODEL   CONSRH CONMOB  IMPACT.I
$
$ Solution for zero bias using Gummel algorithm
SYMB  GUMMEL CARRIERS=0
METHOD ITLIMIT=20
SOLVE   v<electrode>=0
$
$ Switch to Newton for high biases with two carriers
SYMB  NEWTON CARRIERS=2
METHOD ITLIMIT=20
$  Save log file for I-V plot
LOG OUT.F=<filename>
$
$ Now start the solutions and save the solution file using continuation method
SOLVE CONTINUE  electrod=<name> c.vstep= c.vmax= c.imax= c.toler=
SAVE OUT.F=<filename>
$ END THE PROFRAM
```

Fig. 5.1   An example of a MEDICI program

(a) Sample Synopsys TCAD PROGRAM for current-voltage (I/V) characteristics at default room temperature (RT)=300K
* I/V of Non-irradiated p^+n Si strip detector with the two guard rings (Bias GR-0V) for HL-LHC.
# INPUT FILE for
# DC analysis at 500 V

$ Electrode ---*

```
Electrode{
  { name="Pcontact"    voltage=0 }
  { name="Ncontact"    voltage=0 }
  { name="guardring1"  voltage=0.0 }
  { name="guardring2"  voltage=0.0 }
}
```

$ File Description---*

```
File{
  Grid   = "ajau_msh.tdr"
* For Silicon
  parameter= "models.par"
  parameter= "Oxide.par"
  parameter= "Oxide%Silicon.par"
  parameter= "Aluminum.par"
  Plot   = "ajayu"
  Current = "ajayu"
  Output  = "ajayu"
}
```

$ Plot--*

```
Plot {
eDensity hDensity eCurrent hCurrent
Potential SpaceCharge ElectricField
eMobility hMobility eVelocity hVelocity
eQuasiFermi hQuasiFermi Potential
SRHRecombination Auger AvalancheGeneration
Doping DonorConcentration AcceptorConcentration
Current
surfaceRecombination
eIonIntegral hIonIntegral

}
```

$ Physical models---*

```
Physics{
Temperature=300
Mobility( DopingDep HighFieldsat Enormal )
EffectiveIntrinsicDensity( OldSlotboom )
Recombination ( SRH ( DopingDependence )
eAvalanche (CarrierTempDrive) havalanche (Eparallel) hAvalanche(Okuto))
Recombination ( Auger (WithGeneration))
Recombination ( SRH )
ComputeIonizationIntegrals(WriteAll)


}
```

Fig. 5.2 An example of a Synopsys TCAD program for (**a**) current-voltage and (**b**) capacitance-voltage measurement

```
$ Si/SiO₂ Interface Physics----------------------------------------------------------------- *

Physics (
MaterialInterface ="Silicon/Oxide") {
charge= (Conc=o)
Recombination(surfaceSRH)
}


$ Math ------------------------------------------------------------------ *

Math{
  Extrapolate
  Derivatives
  NewDiscretization
  NotDamped=50
  Iterations=100
BreakATIonIntegral
}

$ Solve---------------------------------------------------------------- *

Solve {

  Poisson
  Coupled{ Poisson Electron Hole }

Quasistationary (
    InitialStep=0.2 Increment=0.01
    MaxStep =0.01 MinStep = 0.01
    Goal{ Name="Ncontact" voltage = 500}
  ){ Coupled {Poisson Electron Hole} }
```

*(b) Sample Synopsys TCAD PROGRAM for capacitance-voltage (C/V) characteristics at default room
temperature (RT)=300K*
* C/V of Non-irradiated p^+n Si strip detector with the two guard rings (Bias GR-0V) for HL-LHC.
# INPUT FILE for
# AC analysis at 10 KHz

```
$ Electrode---------------------------------------------------------- *

Device GD {

Electrode{
  { name="Pcontact"     voltage=0 }
  { name="Ncontact"     voltage=0 }
  { name="guardring1"   voltage=0 }
  { name="guardring2"   voltage=0}

}

$ File----------------------------------------------------------- *

File{

  Grid   = "ajayu_msh.tdr"

* For Silicon
  parameter= "models.par"
  parameter= "Oxide.par"
  parameter= "Oxide%Silicon.par"
  parameter= "Aluminum.par"
```

Fig. 5.2 (continued)

```
Plot   = "ajayu"

}
```

$ Physical models--*

```
Physics{
Temperature=300
Mobility( DopingDep HighFieldsat Enormal )
EffectiveIntrinsicDensity( OldSlotboom )
Recombination ( SRH ( DopingDependence )
eAvalanche (CarrierTempDrive) havalanche (Eparallel) hAvalanche(Okuto))
Recombination ( Auger (WithGeneration))
Recombination ( SRH )
ComputeIonizationIntegrals(WriteAll)

}
```

$ Si/SiO₂ interface--*

```
Physics (
MaterialInterface ="Silicon/Oxide") {
charge= (Conc=1e10)
Recombination(surfaceSRH)
}
```

```
*_____*
```

$ Plot---*

```
Plot {
eDensity hDensity eCurrent hCurrent
Potential SpaceCharge ElectricField
eMobility hMobility eVelocity hVelocity
eQuasiFermi hQuasiFermi Potential
SRHRecombination Auger AvalancheGeneration
Doping DonorConcentration AcceptorConcentration
Current
eIonIntegral hIonIntegral
eTrappedCharge hTrappedCharge
SurfaceRecombination

}
```

```
*_____*
}
```

$ Math---*

```
Math{
   Extrapolate
   RelErrcontrol
   Notdamped=50
   BreakATIonIntegral
   Iterations=20
}
```

$ File--*

```
File {
Output = "ajayu"
ACExtract="ajayu "
}
```

Fig. 5.2 (continued)

$ AC Signal ---*

System
{

GD trans(Pcontact Ncontact guardring1 guardring2)

Vsource_pset Vp   (Pcontact 0) {dc=0}
Vsource_pset Vn   (Ncontact 0) {dc=0}{sine= (0 0.5 0.01meg 0 0)}
Vsource_pset Vgr1 (guardring1 0) {dc=0}
Vsource_pset Vgr2 (guardring2 0) {dc=0}

}

$ Solve--*

Solve {

 Poisson
 Coupled{ Poisson Electron Hole }
 # Ramp Vn and apply ac on Ncontact

Quasistationary (
     InitialStep=0.1 Increment=1.3
     MaxStep =0.5 MinStep = 1.0e-5
     Goal{ Paramater=Vn.dc Voltage= 0}
     ){ Coupled {Poisson Electron Hole}}

Quasistationary (
     InitialStep=0.01 Increment=1.3
     MaxStep =0.04 MinStep = 1.0e-5
     Goal{ Paramater=Vn.dc Voltage= 500}
)

{ ACCoupled (

StartFrequency =1e4 EndFrequency=1e4
NumberOfPoints=1 Decade
Node(Pcontact Ncontact guardring1 guardring2) Exclude (Vp Vn Vgr1 Vgr2)
)

{Poisson Electron Hole}}
 }

Fig. 5.2 (continued)

References

1. TMA TSUPREM-4 V. 1999.4. User manual, 1999.4
2. TMA MEDICI V.2000.4. Users manual, February 2000.4
3. Rafferty, C.S., et al.: Appl. Phys. Lett. **68**(17), 2395 (1996)
4. Huang, R.Y.S., et al.: J. Appl. Phys. **74**(9), 5821 (1993)
5. Watts, J.T.: Surface Mobility Modeling, presented at Computer-Aided Design of I.C. Fabrication Processes. Stanford University (Aug. 3, 1998)
6. Caughy, D.M., Thomas, R.E.: Proc. IEEE. **55**, 2192 (1967)
7. Varga, R.S.: Matrix Iterative Analysis. Prentice Hall, Englewood Cliffs, NJ (1962)
8. Selberher, S.: Analysis and Simulation of Semiconductor Devices. Springer, Wien (1984)

9. Chatterji, S., Ranjan, K., Bhardwaj, A., Namrata, Srivastava, A.K., Shivpuri, R.K.: Annealing behaviour of boron implanted defects in Si detector: impact on breakdown performance. Eur. Phys. J. Appl. Phys. **17**, 223 (2002)
10. Chatterji, S., Bhardwaj, A., Ranjan, K., Namrata, A.K.S., Shivpuri, R.K.: Analysis of interstrip capacitance of Si microstrip detector using simulation approach. Solid State Electron. **47**(9), 1491 (2003)
11. Srivastava, A.K., Bhardwaj, A., Ranjan, K., Namrata, Chatterji, S., Shivpuri, R.K.: Two dimensional breakdown voltage analysis and optimal design of Si microstrip detector passivated by dielectric. Semicond. Sci. Technol. **17**, 427 (2002)
12. Synopsys Inc., TCAD software. http://www.synopsys.com/Tools/TCAD/DeviceSimulation/

Chapter 6
Development of Radiation Hard p$^+$n Si Pixel Sensors for the European XFEL

6.1 Introduction

RESEARCH at the European XFEL (X-ray Free Electron Laser) requires pixel sensors with unprecedented performance: Doses of up to 1 GGy 12 keV X-rays and a dynamic range from 1 to 10^5 12 keV photons per sub picoseconds pulse and pixel of (200 μm)^2 area [1].

A lot of progress has already been made in order to understand the radiation hardness of silicon sensors for XFEL environment and the important observation has been found that after a few MGy of irradiations, saturation of charge occurs [2] and on this basis the microscopic parameters proposed for the simulation using experimental results [3].

In XFEL environment (no bulk damage for E$_r$ < 300 keV), the surface charges (oxide charges and interface charges) will degrade macroscopic performance of Si sensors: increases of depletion voltage and surface current, change of interstrip capacitance and interstrip resistance. TCAD simulation was used for the understanding of influence on Si microstrip strip sensor from surface charges after 1 and 10 MGy X-ray irradiation. The final goal of the present work is to develop radiation hard Si pixel sensor for Adaptive Gain Integrating Pixel Detector (AGIPD) at XFEL [4].

In this paper, we have shown the detailed comparison between simulations and measurements for non-irradiated and irradiated CMOS capacitor (annealed 60 min 80 °C) and Si microstrip sensor. The p$^+$n Si pixel sensor is proposed within Adaptive Gain Integrating Pixel Detector (AGIPD) collaboration.

6.2 Test Structures Design and Simulation Technique

The CMOS capacitor (see Fig. 6.1) and AC coupled Si strip test structure sensor (see Fig. 6.2) is used for the detailed comparison with simulation and experimental results and produces X-ray radiation damage results. On this basis, the optimal radial

© Springer Nature Switzerland AG 2019
A. K. Srivastava, *Si Detectors and Characterization for HEP and Photon Science Experiment*, https://doi.org/10.1007/978-3-030-19531-1_6

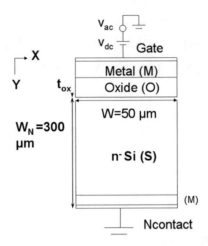

Fig. 6.1 Schematic of the CMOS capacitor for the simulation

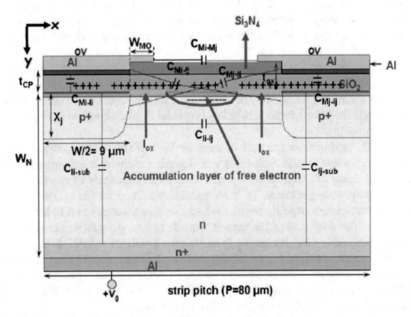

Fig. 6.2 Schematic of 2 strips AC coupled Si sensor for the simulation

hard Si sensor design is proposed using 2-D synopsys TCAD device simulation
[5]. The geometrical and physical parameters for both designs are described in
Tables 6.1, 6.2, 6.3 and 6.4.

Table 6.1 List of geometrical parameters for CMOS Capacitor

S.No.	Geometrical parameters	Values
1.	MOS gate width(W)	10 µm
2.	Diameter	1.5 mm
3.	Full volume of MOS test structure (2-D)	X = 10 µm, Y = 300 µm
4.	Default gate length (L) in 2-D simulation	1 µm

Table 6.2 List of physical and biasing parameters for CMOS capacitor

S.No.	Physical parameters	Values
1.	Oxide thickness (t_{ox})	0.405 µm
2.	Net doping concentration in bulk Si	1.38×10^{12} cm^{-3}
3.	Depth of net doping (d_d) [µm]	20–300 µm
4.	Doping concentration near to Si-SiO$_2$ interface	7.5×10^{11} cm^{-3}
5.	Depth of surface doping profile (d_s) [µm]	0–20 µm
5.	Gate voltage (V_g)	Depends upon N^{fix}_{ox}/V_{fb}, D_{it}
6.	AC voltage (V_{AC})	0.1 V
7.	Frequency (f)	10, 100 kHz

Table 6.3 List of geometrical parameters for AC coupled Si sensor

S.No.	Geometrical parameters	Values
1.	2 strip sensor pitch (P)	80 µm
2.	Width of strip (W)	18 µm
3.	Length of strips (L)	7.8 mm
4.	Default length of strips (L) in 2-D simulation	1 µm
5.	Total area of 98 strips	0.643 cm^2
6.	Full volume of 2 strip sensors (2-D)	X = 80 µm, Y = 282 µm

Table 6.4 List of physical and biasing parameters for AC coupled Si sensor

S.No.	Physical parameters	Values
1.	Coupling oxide + nitride thickness (t_{cp})	200 nm + 50 nm
2.	Field oxide + nitride thickness (t_{ox})	300 nm + 50 nm
3.	Net doping concentration in bulk Si	8.1×10^{11} cm^{-3}
4.	Junction depth (X_j)	1 µm
5.	Thickness of n$^+$ (W_{n+})	1 µm
6.	Thickness of Al strips	1 µm
7.	Width of metal-overhang	1 µm
5.	Bias voltage (V_g)	100 V, 500 V
6.	AC voltage (V_{AC})	0.1 V
7.	Frequency (f)	100 kHz, 1 MHz

6.3   Specification of the p⁺n Si Pixel Sensor for AGIPD

The specs. for the sensors are given below:
    Main specifications

- radiation tolerance: 0 … 1 GGy
- $I_{pixel} < 1$ nA; $I_{tot} < 3$ mA (minimize I_{dark})
- inter-pixel capacitance $C_{pixel} < 0.5$ pF (minimize C_{pixel})
- flatness: <25 mm (minimize flatness)
- minimize dead region (aim for 0.5 mm)
- breakdown voltage >500 V
- interstrip/pixel resistance >20 GΩ

In this note, we have optimized the pixel design (consider like as a strip in present 2-D TCAD simulation) for breakdown voltage V_{BD}, Interstrip/pixel capacitance C_{int}, Surface current I_{ox}, and Interstrip/pixel resistance R_{int} and minimize the region of low electric field for no charge loss.

6.3.1   AGIPD Sensor

The p⁺n Si pixel sensor will be circular bump bonded with usually Indium of diameter (10–25 micron) to the automatic gain switching readout ASIC electronics (see Fig. 6.3). The window will open in the final passivation for bump bonding with each pixel.
    The detailed descriptions of readout ASIC electronics are presented in [4].

Fig. 6.3 Cross-section of one DC coupled p⁺n Si pixel sensor (bump bonded with readout)

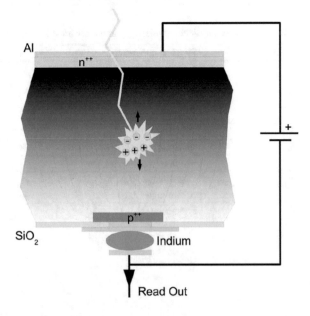

6.4   Physics of Breakdown Voltage V_{BD}, Interstrip Capacitance C_{int}, Surface Current I_{ox}, and Interstrip Resistance R_{int}

In reverse bias p^+n silicon pixel sensor, the voltage at which the current increases sharply is called the avalanche breakdown voltage V_{BD}. The critical electric field for the breakdown is 3×10^5 V/cm and the impact ionization coefficient (α_n, or α_p or both) should be equal to one.

The interstrip capacitance is the creation of series noise in the readout ASIC electronics therefore the sensor design should be carefully optimized. The cross-talk is also measurable and this is calculated by ratio of capacitance of neighbors with total capacitance i.e. sum of detector capacitance and bump bonding capacitance, and charge sensitive preamplifier capacitance. In an array of silicon pixel sensor, the interstrip capacitance is the sum of the all neighboring adjacent pixels and diagonal pixels. A lot of work has been already done to calculate the C_{int} in 3×3 pixel array using analytical methods, HSPICE circuit simulator, and 2-D, 3-D ISE TCAD [6, 7].

In an irradiated silicon pixel sensor by 12 keV X-rays, the flow of dark surface current I_{ox} is also the creation of parallel shot noise in the readout ASIC electronics due to high interface trap density at Si-SiO$_2$ interface of sensor.

The measurement of interstrip resistance R_{int} gives the qualitatively information about spatial resolution.

6.5   Strategy for Simulation

It is known from previous studies [8] on pixel sensor that the C_{int} is strongly influenced by the gap between the pixel and also dark surface current I_{ox} increases with gap. Therefore it is important to optimized the gap for 500 V operation without any avalanche breakdown in order to limit charge explosion effect (plasma of high e/h densities) [9], low C_{int}, dark current (within specs) and $R_{int} > 20$ GΩ.

The p^+n Si pixel sensor for the outer region will be investigated later.

In the first step, we have compared the experimental and simulation result using TCAD device simulation of non-irradiated CMOS capacitor. The fixed oxide charge density (Nox) is used from the experiment. The parameters are determined i.e. interface densities (D_{it}), width of gaussian σ^{rms}_{it}, and energy level E_c-E_{it}, and the calculated capture cross-sections of D_{it} ($\sigma_{eff} = 7 \times 10^{-17}$ cm^2) are used here from the non-irradiated gated diode measurements [Sri]. For non-irradiated CMOS capacitor, it is very difficult to get the microscopic parameters from experiment.

In the next step, we have compared with experimental data on 60 min 80 °C 5 MGy irradiated CMOS capacitor. The fixed oxide charge densities (N_{ox}), interface densities (D_{it}), width of gaussian σ^{rms}_{it}, and energy level E_c-E_{it} is used from the experiment and the calculated capture cross-sections of D_{it} ($\sigma_{eff} = 7 \times 10^{-17}$ cm^2) are used here from the non-irradiated and irradiated gated diode measurements

[Sri[1]]. It is believed that the capture cross-section of D_{it} changes with time and after some time it will attain the pre-damage value [10].

It has been found that the slope of the irradiated C/V$_g$ curve is very sensitive to the interface densities (D_{it}), width of gaussian σ^{rms}_{it}, and energy level E_c-E_{it}.

In order to get good qualitatively agreement with experimental data, we have refine the microscopic parameters specially width of gaussian σ^{rms}_{it} and E_c-E_{it} within ± 0.05 eV (from experimentally measured TDRC, D_{it} versus E_c-E_{it}). It should be noted the taken values are within the experimental error and it the precise accurate estimation of σ^{rms}_{it} is not possible because of the offset problem of the TDRC curve. The model calculations should be carefully performed for the shallower and deeper interface trap to get microscopic parameters using the complex RC network inside the CMOS capacitor because surface damage is a complex business of surface charge effects.

The following two Gaussian interface trap model are used in TCAD device simulation [11].

6.6   Simulation Results

In this section, we have presented the simulation results performed within the framework of the AGIPD collaboration for the development of adaptive gain integrated hybrid silicon pixel detectors at XFEL.

6.7   Simulation Results of Test Structure

The doping profiles of CMOS capacitors are extracted from the experiment (C-V and G-V) and verified using analytical expressions [Sri]. The knowledge of accurate doping profiles is important for C-V modeling of CMOS capacitor.

In the first set of simulation results, we have shown the detailed comparison of experimental results and simulation results of non-irradiated CMOS capacitor. The results are shown in Fig. 6.4. The following microscopic parameters are extracted from TCAD: $D_{it} = 2.66 \times 10^{11}$ cm^{-2} eV^{-1}, E_c-$E_{it} = 0.4$ eV, $\sigma^{rms}_{it} = 0.1$ eV. It can be seen that the capacitance is well described qualitatively in all regions at 10, and 100 kHz. The flatband shift is also observed with increasing frequency from 10 to 100 kHz due to high value of interface trap density. It is also observed that there is also change of capacitance in strong accumulation region of CMOS capacitor. This is due to the uncorrected series resistance of undepleted bulk of Si [Sri] or may be due to unwanted dielectric loosy layer formed and increases at Si-SiO$_2$ interface with

---

[1]Sri is referenced by [11].

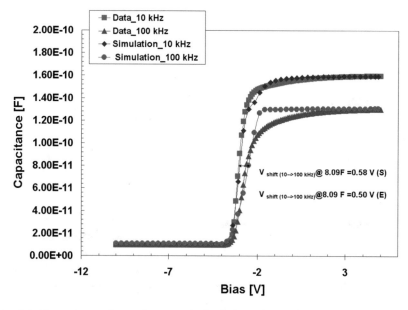

Fig. 6.4 Comparison of experiment and simulation of capacitance–voltage (C/V_g) for non-irradiated CMOS capacitor at 1 and 100 kHz frequency

increasing frequency and thus gives an unobvious effect i.e. decreases of oxide capacitance in strong accumulation [Sri].

The comparison of experimental and simulated conductance–voltage (G-V) characteristics of non-irradiated CMOS capacitor is shown in Fig. 6.5(a and b) at 10 and 100 kHz.

Here we have also explained the importance of G-V characteristics for surface charge effects. The three regions (G_{acc}, G_{peak} and G_{inv}) of G-V curve are already described in our first preliminary note on non-irradiated CMOS capacitor [Sri]. Here I have discussed our observation on the basis of simulation results. G_{acc} is depend on the series resistance of the undepleted bulk of Si and G_{peak} gives the information about the charge carrier distribution at Si-SiO$_2$ interface. Whereas, G_{inv} is strongly depend on the Temperature of measurement and also doping in the space charge region. It is observed that G_{acc}, G_{peak} and G_{inv} increases with frequency. The good agreements in G_{acc}, G_{peak} and G_{inv} with experimental results are only possible in the presence of accurate doping profile and microscopic parameters. This is strongly verified our observed microscopic parameters for non-irradiated CMOS capacitor.

It is noted that the conductance is well described qualitaveley in the accumulation, depletion and inversion region and the peak range problem occurs due to sharply decrease of conductance in the accumulation region at 10 kHz. The reason is possibly due to the measurements because the previous measurement on gated diode shows the simulated G-V curve at 10 kHz [Sri].

Now, we have compared the experimental result of 60 min 80 °C annealed 5 MGy irradiated CMOS capacitor with simulation. The date was taken from TDRC measurement (see Fig. 6.6a). The strategy used the same as per non-irradiated CMOS

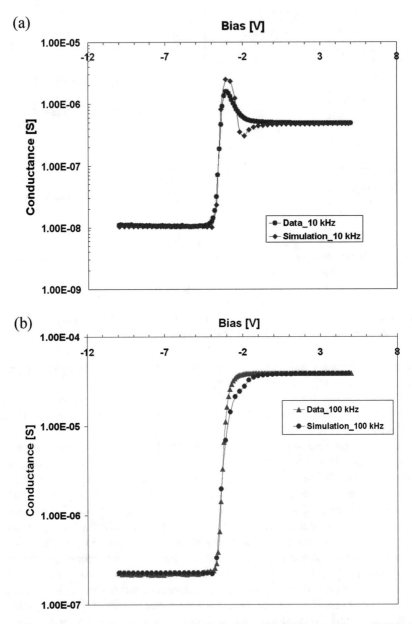

Fig. 6.5 Comparison of experiment and simulation of conductance-voltage (G/V$_g$) for non-irradiated CMOS capacitor at (**a**) 1 and (**b**) 100 kHz frequency

capacitor: firstly to compare with simulation and then refine E_c-E_{it}, σ^{rms}_{it} (close to experimental value) in order to get good agreement in C/V and G/V (see Fig. 6.6b and c).

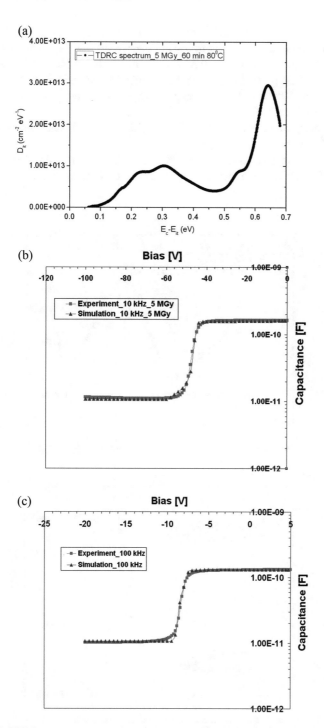

Fig. 6.6 (**a**) TDRC measurement result and (**b**) comparisons of experiment and simulation of capacitance-voltage (C/V$_g$) for 5 MGy irradiated CMOS capacitor after 60 min 80 °C at 10 kHz and (**c**) 100 kHz frequency

For 5 MGy irradiated CMOS capacitor, the C/V curve is in good agreement with experimental results for the following parameters (see Fig. 6.6b and c): microscopic defect: Acceptor, E_c-E_{it}: 0.35 eV, Gaussian distribution σ^{rms}_{it}: 0.06791 eV and Donor, E_c-E_{it}: 0.60 eV, Gaussian distribution σ^{rms}_{it}: 0.0065 eV with $N_{ox} = 2.3 \times 10^{12}$ cm^{-2} and Acceptor, $D_{it} = 1 \times 10^{13}$ cm^{-2} eV^{-1}, Donor, $D_{it} = 4 \times 10^{13}$ cm^{-2} eV^{-1}.

The comparison of experiment with simulation for G-V characteristics is shown in Fig. 6.7a and b.

The G-V curve (G_{peak}) is qualitatively good agreement in simulation at 10 and 100 kHz frequency. In experimental measured G-V curve, G_{inv} is increasing at

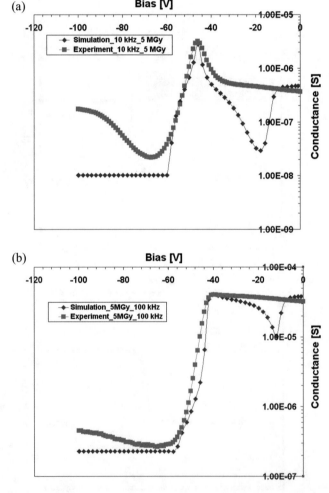

Fig. 6.7 Comparison of experiment and simulation of capacitance-voltage (G/V$_g$) for 5 MGy irradiated CMOS capacitor after 60 min 80 °C at (**a**) 1 and (**b**) 100 kHz frequency

10 and 100 kHz frequency, the reason is still not clear. In simulation, it seems to be constant as expected like from non-irradiated CMOS capacitor. In simulated G-V curve, in accumulation region of CMOS capacitor, G_{acc} is attaining the same positions as expected at 10 and 100 kHz like from non-irradiated CMOS capacitor but in experimental curve, it is not observed.

Finally, the results are cross-checked with analytical expressions [Sri].

6.8   Simulation Results of Test Structure AC Coupled p⁺n Silicon Strip Sensor

In the framework of AGIPD collaboration, 98 Si strip sensors were irradiated by 10 keV X-ray doses of 1 MGy and 10 MGy and annealed up to 60 min 80 °C. The simulator is used to compare the experimental results of dark current, total back-plane capacitance, and interstrip capacitance with simulation results and present X-ray radiation damage results. In order to check accuracy of simulator and validation of out design, the simulation results are verified using analytical expressions for leakage current, back-plane capacitance, interstrip capacitance for ideal case when there is no fixed charge in the oxide and at Si-SiO$_2$ interface. The good agreement in leakage current but for back-plane, interstrip, and total detector capacitance within the agreement of 10–23–30% with analytical expressions for the w/p = 0.2 [12]. It is important to mention that the 100% agreement is achieved for w/p = 0.6 (see Sect. 6.3).

The microscopic defects (N_{ox}, D_{it}) for 1 MGy and 10 MGy were taken from experiment [3] and E_c-E_{it} and Gaussian distribution σ^{rms}_{it} for both shallower and deeper interface trap from previous simulation. There is no comparison with simulation has been made for 1 and 10 MGy with C/V and G/V because of unavailability of data.

The following parameters: microscopic defect: Acceptor, E_c-E_{it}: 0.35 eV, Gaussian distribution σ^{rms}_{it}: 0.06791 eV and Donor, E_c-E_{it}: 0.60 eV, Gaussian distribution σ^{rms}_{it}: 0.0065 eV with $N_{ox} = 2.08 \times 10^{12}$ cm^{-2} and Acceptor, $D_{it} = 4.2 \times 10^{12}$ cm^{-2} eV^{-1}, Acceptor, $D_{it} = 1 \times 10^{13}$ cm^{-2} eV^{-1} were used only for 1 MGy. The capture cross-sections of both interface trap (shallower and deeper), $\sigma_{eff} = 7 \times 10^{-17}$ cm^2 were used.

In 1 MGy irradiated sensor, the effect of Nox and D_{it} on depletion behavior is clearly shown in Fig. 6.8. In the presence of N_{ox} (no D_{it}), the depletion merge (V_{merge}) at around 5 V but in the presence of N_{ox} + D_{it} both, V_{merge} improved upto few volts (1–2 V). V_{merge} can be seen in 1/C^2 versus bias characterstics of sensor.

The electric field minimum has been found at Y = 10 μm (see Fig. 6.9a and b). This can be exempted from any loss of charge. This is because of no electric field lines are passing through the small volume of the sensor.

It is shown in Fig. 6.10a, N_{ox} increases (0 to 1×10^{11} cm^{-2}) the full depletion voltage up to 10–15 V because of delayed depletion depth and for further increase in N_{ox}, V_{FD} saturate.

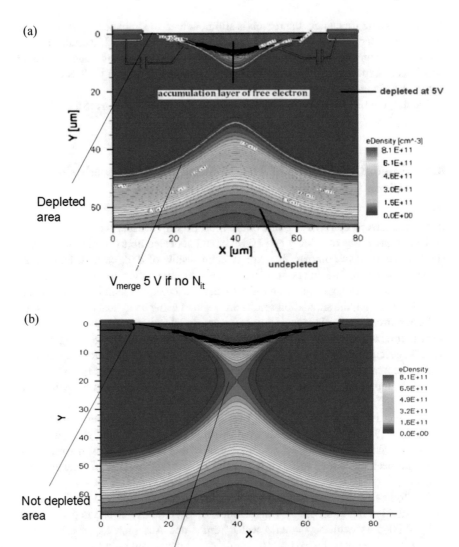

Fig. 6.8 e-density inside AC coupled Si strip sensor at 5 V (**a**) Nox ~1 MGy, $D_{it} = 0$ (**b**) Nox and D_{it} both ~1 MGy irradiated doses

The good agreement in experimental data and simulation is observed in total back-plane capacitance of non-irradiated strip sensor for $N_{ox} = 2 \times 10^{10} \text{ cm}^{-2}$. The effect of ($N_{ox} + D_{it}$) in irradiated sensors on full depletion voltage can be seen in Fig. 6.10b. In 1, 10 MGy irradiated sensor ($N_{ox} + D_{it}$), there is a 1–2 V of full depletion voltage changes and it is due to change of an effective oxide charges at Si-SiO₂ interface.

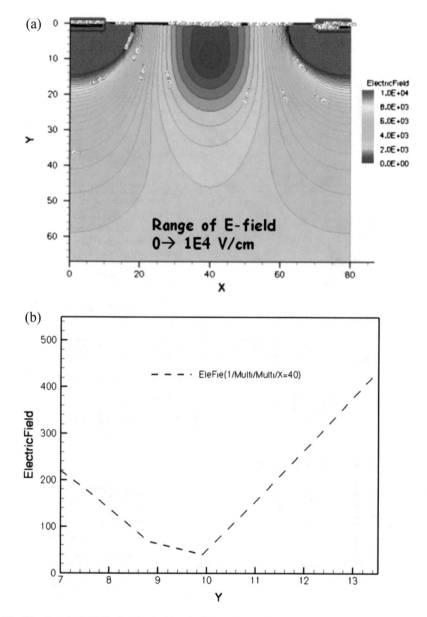

Fig. 6.9 Electric field distribution inside 1 MGy irradiated AC coupled Si strip sensor at 100 V (**a**) E-field (range of E-field is 0 to 1×10^4 V/cm.) (**b**) Electric field versus depth at X = 40 μm

It has been found that the generation/recombination surface current increases with non-implanted surface depleted area and also with voltage (see Fig. 6.11). There is a disagreement in expected experimental current of 9 μA/cm². This is because of the high value of the effective capture cross-section of the shallower and deeper traps. In

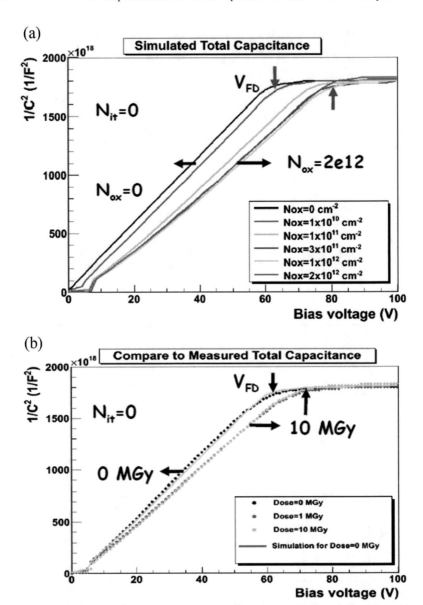

Fig. 6.10 (a) $1/C^2$ versus bias voltage for different fixed oxide charge ($D_{it} = 0$) at 100 kHz frequency (b) $1/C^2$ versus bias voltage for different irradiated sensor. Comparison of experiment with non-irradiated sensor for $N_{ox} = 2 \times 10^{10}$ cm^{-2} at 100 kHz frequency

order to make a good agreement in simulated surface current at every voltage with experimental data, we have tune the effective capture cross-section of interface traps.

Firstly, it is important to see the effect of boundary conditions on surface current in order to verify the suitable boundary condition for the present simulation approach.

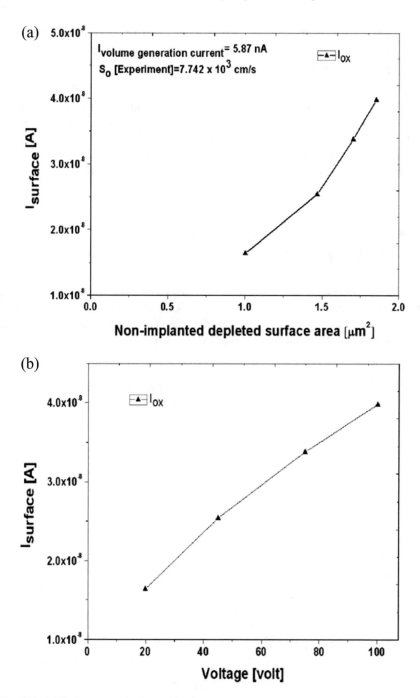

Fig. 6.11 (**a**) Surface generation/recombination current versus non-implanted surface depleted area and (**b**) voltage in 1 MGy irradiated sensor

6.9   Effect on Surface Current in Gate Boundary Condition and Neumann Boundary Condition in 5 MGy Irradiated (60 °C 80 min Annealed)

When gate boundary conditions used it has been found that the surface current increases with voltage but saturate up to 60 V whereas in experimental the surface current continuously increases with voltage (see Fig. 6.12).

Therefore, Neumann boundary condition is an appropriate boundary condition for the comparison of simulated surface current with experimental data.

6.10   Comparison of Simulated Surface Current with Experimental Data in an Irradiated Sensor

In order to make good agreement in simulation surface current and experimental data, we have tune the capture cross-sections of both interface trap and for the value of 2.75×10^{-15} cm^2. The surface current of 9 µA/cm^2 is in good agreement with the measurement results on gated diode (see Fig. 6.13a and b) [2].

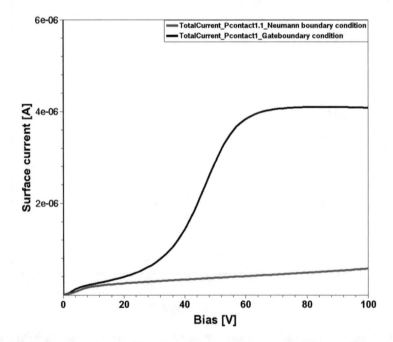

Fig. 6.12 Surface current versus bias voltage for gate and Neumann boundary condition

(a)

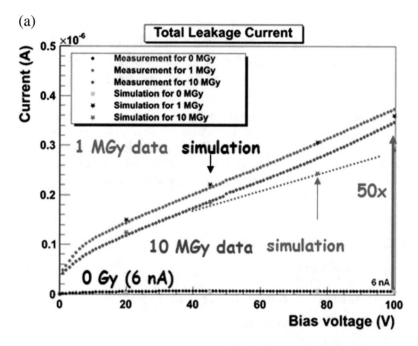

(b)

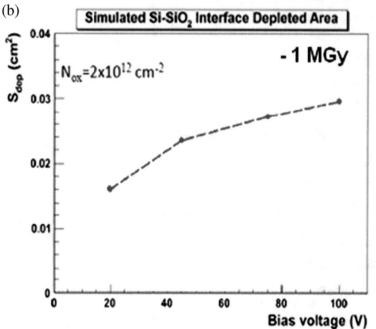

Fig. 6.13 Comparison of (**a**) simulated surface current and experiment for 1 and 10 MGy irradiated sensor and (**b**) total normalized depleted area for 98 strip AC coupled sensor

6.11   Model Parameters and Comparison of Simulated Total Back-Plane Capacitance (Serial and Parallel) and Interstrip Capacitance (Serial and Parallel) with Experimental Data in an Irradiated Sensor

There is model calculation performed which is based on the experiment result and analytical expressions given in books and on this basis; two level interface trap microscopic parameters for 1, 5 and 10 MGy irradiated doses are derived.

The parameters are shown below. The value of N_{ox} below is shown for immediately after irradiation where as D_{it} and other parameters are shown after annealing.

Parameters for 1, 5 and 10 MGy:

1 MGy: $N_{ox} = 3.3 \times 10^{12}$ cm^{-2}

- E_c-$E_{it} = 0.354$ eV (acceptor)

  - Gauss with rms $= 0.1016$ eV
  - $\sigma_{eff} = 1 \times 10^{-16}$ cm^2
  - D_{it} (0.354 eV) $= 6.438 \times 10^{12}$ cm^{-2} eV^{-1}

- E_c-$E_{it} = 0.636$ eV (acceptor)

  - Gauss with rms $= 0.045$ eV
  - $\sigma_{eff} = 4 \times 10^{-15}$ cm^2
  - D_{it} (0.636 eV) $= 7.118 \times 10^{12}$ cm^{-2} eV^{-1}

5 MGy: $N_{ox} = 3.1 \times 10^{12}$ cm^{-2}

- E_c-$E_{it} = 0.354$ eV (acceptor)

  - Gauss with rms $= 0.1016$ eV
  - $\sigma_{eff} = 1 \times 10^{-16}$ cm^2
  - D_{it} (0.354 eV) $= 5.861 \times 10^{12}$ cm^{-2} eV^{-1}

- E_c-$E_{it} = 0.636$ eV (acceptor)

  - Gauss with rms $= 0.045$ eV
  - $\sigma_{eff} = 4 \times 10^{-15}$ cm^2
  - D_{it} (0.636 eV) $= 6.728 \times 10^{12}$ cm^{-2} eV^{-1}

10 MGy: $N_{ox} = 3.0 \times 10^{12}$ cm^{-2}

- E_c-$E_{it} = 0.354$ eV (acceptor)

  - Gauss with rms $= 0.1016$ eV
  - $\sigma_{eff} = 1 \times 10^{-16}$ cm^2
  - D_{it} (0.354 eV) $= 5.024 \times 10^{12}$ cm^{-2} eV^{-1}

- E_c-$E_{it} = 0.636$ eV (acceptor)

  - Gauss with rms $= 0.045$ eV
  - $\sigma_{eff} = 4 \times 10^{-15}$ cm^2
  - D_{it} (0.636 eV) $= 6.337 \times 10^{12}$ cm^{-2} eV^{-1}

Therefore, in simulation we have used the $N_{ox} = 2.3 \times 10^{12}$, 2.1×10^{12}, 2.0×10^{12} cm^{-2} for 1 MGy, 5 MGy and 10 MGy irradiated detectors (60 min 80 °C).

It has been found that in Fig. 6.14a there is a little decrease of non-implanted depleted surface area observed with increases of doses and thus we can say surface current will also little decrease, it can be seen that the non-implanted depleted surface area decreases with increasing N_{ox} for fixed D_{it} for 5 MGy irradiated sensors.

It is well known that the capacitance can be measured in the parallel and serial mode. The serial mode is well accepted because of the appropriate model for simple pn junction. Here we have shown the results for both modes (see Fig. 6.15), it has been found that because of series resistance effect at different frequency (100 kHz) and at 1 MHz, the effect of series resistance will be very less, serial total back plane and interstrip capacitance capacitance is in qualitatively good agreement with simulated result.

In order to make it in good agreement with experimental data on 1 MGy and 10 MGy, the following $N_{ox} = 1.9 \times 10^{12}$ cm^{-2} for 1 MGy and $N_{ox} = 1.55 \times 10^{12}$ cm^{-2} for 10 MGy are extracted from simulation for same D_{it}.

This is in good agreement with our observation on annealed sensors and also that can be accepted because from the experimental measurements, it is difficult to get exact value of N_{ox} in the presence of C/V hysteresis effect in CMOS capacitor (due to N_{ox} mobile charge).

It should be noted that the exact information of $D_{it} + N_{ox}$ is important for the comparison with interstrip capacitance and for total back-plane capacitance, D_{it} and knowledge of accurate doping profile (uniform/non-uniform) is required.

6.12   Simulation Results of p⁺n Silicon Pixel Sensor for AGIPD

In the frame work of the AGIPD collaboration, 200×200 μm^2 of the pixel size is taken as a first consideration. This present simulation approach is optimized the design parameters for the higher and stable performance of the sensors (consider like as a strip for capacitance and resistance in 2D simulation) in terms of low interstip capacitance, dark current within the specs, and acceptable interstrip resistance in the XFEL environment.

In Fig. 6.16, the schematic of the cross-section of the two pixel sensor is shown. The pixel is made on 3–5 kΩ-cm high resistivity n-type FZ Si substrate material of thickness 500 μm and other process parameters are taken in order to keep in mind that there will no influence on the electric field up to 1500 V bias in the presence of surface charge variations and that will protect the sensor from avalanche breakdown. Firstly, we have done two dimensional simulation for an ideal sensor for the leakage current, back-plane capacitance, and interstrip and total back-plane capacitance. The analytical calculation performed for an ideal sensor and the simulation results for the

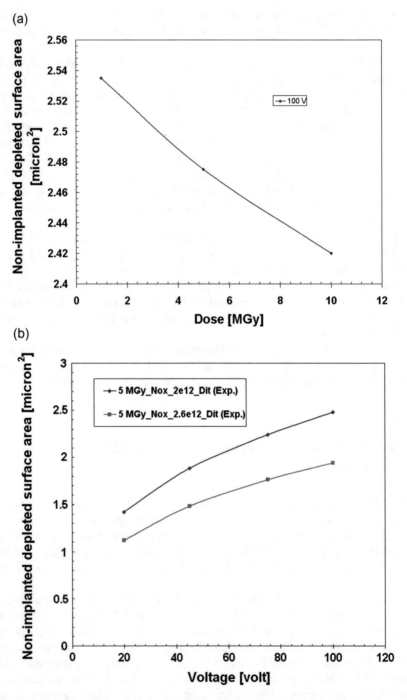

Fig. 6.14 (**a**) Change of non-implanted depleted surface are with irradiated doses. (**b**) Change of non-implanted depleted surface area with voltage with N_{ox} is as running parameter

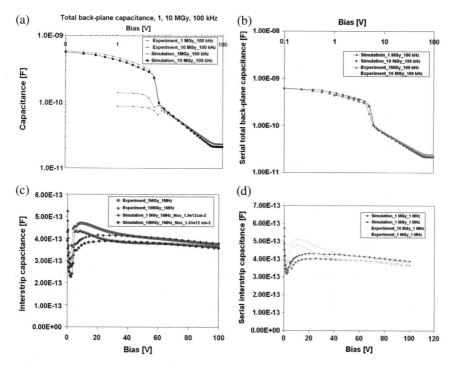

Fig. 6.15 Total back-plane capacitance versus bias voltage for 1 MGy and 10 MGy irradiated and annealed (60 min 80 °C) at 100 kHz (**a**) parallel (**b**) serial and interstrip capacitances versus voltage for 1 MGy and 10 MGy irradiated and annealed (60 min 80 °C) at 1 MHz (**c**) parallel (**d**) serial

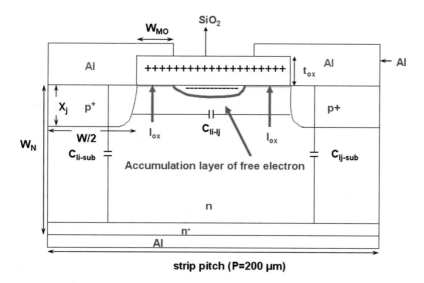

Fig. 6.16 Schematic of the two pixel sensors used in the present simulation work

leakage current, $C_{back-plane}$, C_{int}, and total detector capacitance is in 100% good agreement with theoretical calculation.

The leakage current of 2.56 pA and full depletion voltage of 197 V is found for non-irradiated 200 × 200 μm^2 pixel.

Figure 6.17 show the I-V curve and C-V for the non-irradiated sensor ($N_{ox} = 3 \times 10^{11}$ cm^{-2}) to get the full depletion voltage of the detector.

The influence of the gaps on the pixel dark surface current is shown in Fig. 6.18a for 5 MGy irradiated sensor (annealed 60 min 80 °C).

It is found that the dark surface current increases with gap this is because of the increases of the non-implanted depleted surface area (see Fig. 6.18b). The estimated dark current at 500 V for 80 micron gap is 9 μA/cm^2.

It is found that the peak amplitude of the electric field at curvature of junction increases (<200 kV/cm) with gap size as expected (see Fig. 6.19a) and there is a minimum electric field region at the center of the sensor around 10 micron depth from the Si-SiO^2 interface but there is no charge loss expected because of no files lines passing through this area (see Fig. 6.19b).

The interstrip capacitance decreases with increases gap for non-irradiated and irradiated pixel sensor and the minimum interstrip capacitance is obtained for 80 μm gap (see Fig. 6.20). The estimated value of the C_{int} at full depletion for 200 × 200 μm^2 (sensor pixel array) is 110 fF and the acceptable dark surface current at full depletion is 100 pA for 5 MGy irradiated sensor (annealed 60 min 80 °C).

In Fig. 6.21, I have shown the interstrip resistance versus gap gap for non-irradiated and irradiated pixel sensor (annealed 60 min 80 °C). It is clear that the R_{int} decreases with increases gap and R_{int} for 5 MGy dose is less than the non-irradiated sensor. For 80 μm gap, $R_{int} = 22$ GΩ obtained for 5 MGy irradiated sensor (annealed 60 min 80 °C).

6.13 Outlook

In the next step, the aim is to simulate the 3D geometry of the sensor pixel array for the capacitance calculations and compare the simulated results with analytical expressions. The behaviour of C_{int} versus voltage curve for different design with different gaps in the presence of $N_{ox} + D_{it}$ is always question and in the next work, I will propose the two strip lumped model in order to explain the shape of the curve. The design for the outer region of the pixel sensor will be also performed.

6.14 Conclusion

The detailed comparison of simulation is made with the experimental data on different test structures design and a good qualitatively understanding is obtained. On this basis, I have proposed the radiation hard design of p$^+$n pixel sensor for

(a)

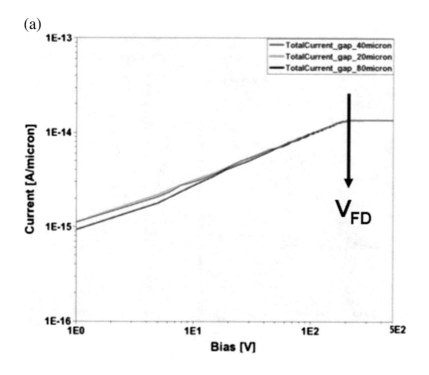

(b)

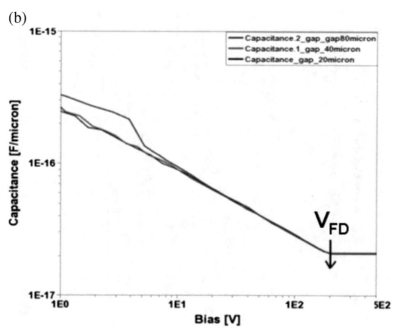

Fig. 6.17 I-V curve and C-V for the non-irradiated sensor ($N_{ox} = 3 \times 10^{11}$ cm^{-2})

(a)

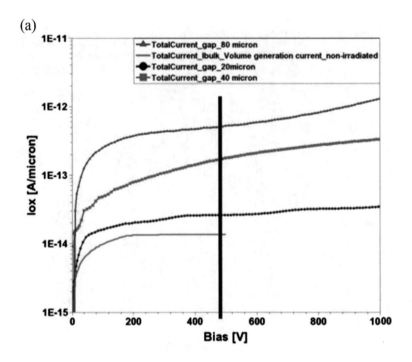

(b)

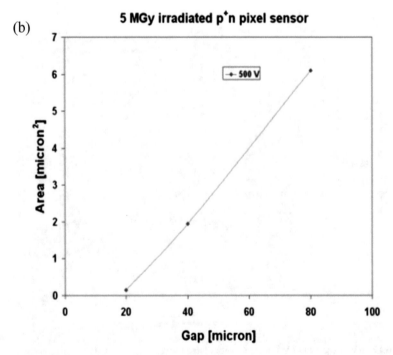

Fig. 6.18 (**a**) Surface current versus bias voltage ($V_{bias} = 500$ V). The gaps between the pixels are as running parameters. (**b**) Non-implanted depleted area versus gap for 5 MGy irradiated sensor (annealed 60 min 80 °C)

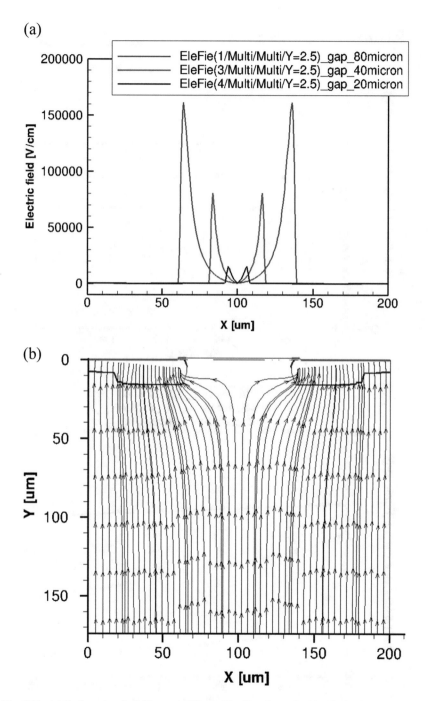

Fig. 6.19 (**a**) Surface electric field versus distance X at Y = 6 μm. (**b**) Electric field lines inside the sensor at 500 V

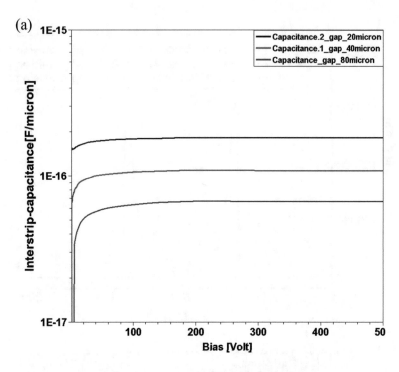

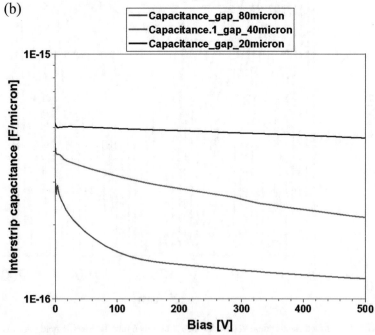

Fig. 6.20 (**a**) Interstrip capacitance versus bias voltage for non-irradiated (**b**) 5 MGy irradiated

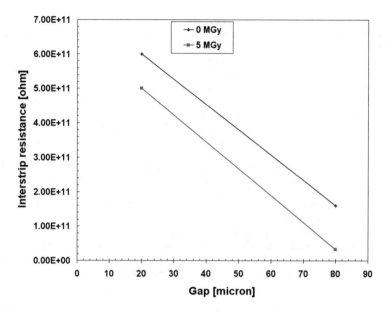

Fig. 6.21 Interstrip resistance versus gap for non-irradiated and 5 MGy irradiated sensor (annealed 60 min 80 °C) at 500 V

AGIPD with 80 micron gap for $200 \times 200 \ \mu m^2$ (sensor pixel array) to reduce the detector load noise (0.21 pF including contribution from bump bonding and automatic gain switching preamplifier load) at the readout of the ASIC electronics that will work up to 500 V with safety margin of 1000 V. In this design, there is no chance to tunnel the electron to the oxide so for ASIC electronics; there will no problem of gate leakage current.

6.15   Acknowledgement

This work is partly funded by the Helmholtz Alliance "Physics at the Terascale" and the German ministry for education and research (BMBF).

References

1. The European X-Ray Laser Project *XFEL*, http://xfel.desy.de/
2. Fretwurst, E., et al.: Study of the Radiation Hardness of Silicon Sensors for the XFEL, poster presented at IEEE NSS 2008, Dresden, Germany, Conference record N30–400
3. Zhang, J., et al.: Study of high-dose X-ray radiation damage of Si sensors. Nucl. Instr. Methods Phys. Res. A. **732**, 117–121 (2013)

4. Göttlicher, P., et al. (AGIPD collaboration): The Adaptive Gain Integrating Pixel Detector (AGIPD): A Detector for the European XFEL: Development and Status, poster presented at IEEE NSS 2009, Orlando, Florida, USA, N25–239
5. Synopsys, Inc., TCAD software. http://www.synopsys.com/Tools/TCAD/DeviceSimulation/
6. Gorfine, G., Hoeferkamp, M., Santistevan, G., Seidel, S.: Capacitance of silicon pixels. Nucl. Instr. Methods Phys. Res. A. **460**, 336–351 (2001)
7. Cerdeira, A., Estrada, M.: Analytical expressions for the calculation of pixel detector capacitances. IEEE Trans. Nucl. Sci. **44**(1), 63–66 (1997)
8. Jens WÜstenfeld: Characterization of ionization induced surface effects for the optimization of silicon detectors for particle physics applications. Ph.D. thesis, University of Dortmund, June 2001
9. Becker, J., Eckstein, D., Klanner, R., Steinbrück, G., on behalf of the AGIPD Consortium: Impact of plasma effects on the performance of silicon sensors at an X-ray FEL. Nucl. Instr. Methods Phys. Res. A. **615**(2), 230–236 (2010)
10. Ma, T.P.: Generation and transformation of interface traps in MOS structures. Microelectron. Eng. **22**, 197–200 (1993)
11. Srivastava, A.K., Fretwurst, E., Klanner, R., Perrey, H.: Analysis of electrical characteristics of gated diodes for the XFEL experiment. DESY internal note (within AGIPD collaboration)
12. Srivastava, A.K., Fretwurst, E., Klanner, R.: Numerical modelling of the frequency behaviour of the irradiated MOS test structure. DESY Internal note (within AGIPD collaboration)

Chapter 7
Analysis & Optimal Design of Radiation Hard p$^+$n Si Pixel Detector for the Next generation Photon Science Experiments

7.1 Introduction

Within the AGIPD collaboration, the R & D efforts on X-ray radiation damage study for the development of radiation hard pixel sensors are performed at the Institute for Experimental Physics, University of Hamburg, Germany. The radiation dose of up to 1 GGy, up to 10^5 photons of 12 keV hitting a 200 μm × 200 μm pixel within 100 fs and a time distance of 222 ns between photon pulses are particular challenges at the European XFEL [1]. Test structures, e.g. CMOS capacitors and gated diodes, fabricated by CiS, Erfurt, Germany have been irradiated with synchrotron radiation white light source at DESY DORIS III. Capacitance-voltage (C/V), conductance-voltage (G/V), current-voltage (I/V) and Thermally Depolarization Relaxation Current (TDRC) measurements have been performed. From these measurements oxide charge densities (N^{fix}_{ox}) and interface densities (D_{it}), capture cross-sections of D_{it} (σ_{eff}), width of Gaussian σ^{rms}_{it}, and energy level E_c-E_{it} have been extracted [2] and implemented into the semiconductor device simulation program Synopsys TCAD [3]. This experience is used to design radiation tolerant p$^+$n silicon pixel sensors.

A lot of simulation efforts have been performed on sensor designing including guard rings and without guard rings and detailed comparison of experimental data and simulation has been made. A good qualitatively agreement between the experimental measurements on irradiated CMOS capacitors, gated diodes, and segmented Si strip sensors [4] and simulation results obtained. We have also performed simulation on n in n sensor design as a first design idea for the AGIPD detector [5].

J. Becker [6] observed that the e-h plasma effect (charge explosion effect) in non-irradiated sensors for the soft XFEL (1 keV, attenuation length of 3 μm) and hard XFEL (12 keV, attenuation length of 230 μm) and suggest that >300 V operating voltage is sufficient in order to eliminate the effect of e-h plasma on change collection time and 60 ns is the maximum charge collection time to collect 95% of produced e-h pairs for 450 μm thick PSI p$^+$n Si strip sensor.

© Springer Nature Switzerland AG 2019

A. K. Srivastava, *Si Detectors and Characterization for HEP and Photon Science Experiment*, https://doi.org/10.1007/978-3-030-19531-1_7

The following implications of the investigations of the surface radiation damage studies for the design of the AGIPD sensor can be summarized. X-ray induced oxide charge and interface trap densities increase for doses up to few MGy and then saturate or even decrease; typical densities are of the order of 10^{13} cm^2 [7]; their introduction rate in a p^+n sensor is somewhat higher if the sensor is irradiated under bias,

– oxide charges and interface traps anneal already at room temperature; the annealing time constants for different interface traps differ by many orders of magnitude,
– the measured radiation behaviour of the dark current, depletion voltage and inter-strip capacitances of segmented sensors can be understood via simulations on the basis of the parameters derived from irradiated test structures,
– electron accumulation layers under the oxide between pixels in p^+n sensor should be minimized, as they result in an increase of the inter-pixel capacitance, in signal losses [3] and possibly instabilities; this requires that sensors should be operated at high bias voltages,
– the area of the Si-SiO$_2$ interface exposed to an electric field should be minimized, as it is the main source of dark current in irradiated sensors,
– the studies show that it should be possible to build both p^+n and n^+n sensors which stand radiation doses up to 1 GGy.

The aim is to design the surface radiation hard Si pixel sensor for AGIPD and some following things is important to note that like oxide changes density varies with the thickness of oxide (tox$^{1.4}$) so it is necessary to keep small oxide thickness for less oxide charge trapping [8]. It is reported that gate boundary conditions can be used for the moisture cases especially for the comparison of the experimental data and TCAD simulation [9]. But, for real case experiment, the NBC (von Neumann boundary conditions) is an appropriate boundary condition for the simulation.

The European XFEL experiment will perform in the dry vacuum environment and for dry case, NBC (von Neumann boundary conditions) will be applied on the outer edge of the oxide surface of the sensors in TCAD simulation [9]. The detector designer and application scientist has a chance to used the real boundary conditions in order to simulate the radiation hard design of the pixel sensors.

The change looses in Si sensors has been already observed in the electron accumulation layer during the transportation of the charge carriers (e/h) produced by an ionizing radiations (for e.g., X-rays) and thus degradation in the charge collection efficiency [10, 11]. In this paper, the following technological important observations are suggested for the designing of radiation hard sensor and we propose two surface radiation hard methods for the future imaging experiments in order to reduce the effect of noise (interpixel capacitances) on ASIC electronics and increase of charge collection efficiency.

This paper is organized in this way, in Sect. 7.2 below test structure designs are summarized. Simulation procedures are presented in Sect. 7.3. Simulation result and discussions are covered in Sect. 7.4. Conclusions are eventually drawn in Sect. 7.5.

7.2   Test Structure Designs

Figure 7.1 shows three one-strip subset of the structure analyzed in the present simulation. With reference to the figure, X_J is the vertical junction depth, W_N is the n-layer thickness, W_{MO} is the width of metal overhang width.

For simulation purposes, we have considered 500 μm thick phosphorus-doped n-type Fz Si wafers, (100) oriented, with resistivity of about 5 kΩ-cm which corresponds to a uniform substrate doping concentration (N_B) of about 1.0×10^{12} cm^{-3}. The back is uniformly n^+ implanted ($W_{n+} = 1$ μm) and metallized to create the ohmic contact. Depletion is attained by positively biasing the back ohmic contact.

The p^+-n junction is assumed to be cylindrical at its edge with a lateral diffusion depth at the curvature of the p-region equal to 0.8 times the vertical junction depth ($X_J = 1$ μm) and a Gaussian profile of the p^+ doping (5×10^{19} cm^{-3} at surface and 1×10^{15} cm^{-3} at the junction is used. The value of the oxide and nitride thickness (t_{ox}) over the Si substrate is kept fixed at 0.35 μm (0.3 μm oxide+0.05 μm nitride) in all simulations. The breakdown voltage of the proposed structures shown in Fig. 7.1 is computed using a 2-D device simulator Synopsys TCAD. The specifications of the AGIPD sensors are given below in Table 7.1.

In the present paper, a simulated standard p^+n structure is compared (Fig. 7.1a) with the two radiations hard p^+n sensor designs (Fig. 7.1b—two strips sensor design with 10 μm p-stop dose (1.125×10^{15} cm^{-2}), Fig. 7.1c—two strips sensor design with a thin 2 μm layer on top of the oxide surface). In three pixel sensor designs, 30 μm pixel gap in between the pixels (strips like) with different overhang width are used.

7.3   Simulation Procedure

We use the Synopsys-TCAD device simulator. It solves Poisson's equation, the continuity, energy balance and lattice heat equations for holes and electrons. These equations describe the static and dynamic behaviour of carriers in semiconductors under the influence of external fields. The physical models used were Shockley–Read–Hall and Auger recombination, and field-dependent mobility. The impact ionization model is used to calculate the breakdown voltage of the sensor.

We use the following boundary conditions, constant potentials (Dirichlet at the contacts) and external field (Neumann at the SiO_2 surface and symmetry at the boundaries inside the sensor).

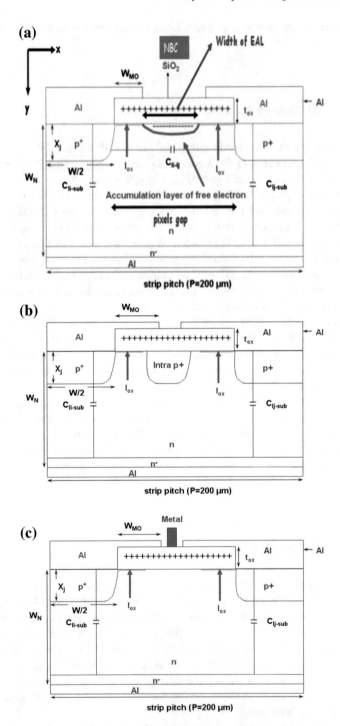

Fig. 7.1 (**a**) A schematic of the p$^+$n Si pixel (strips used) sensors (**b**) a schematic of the p$^+$n Si pixel (strips used) sensors with a p-stop between the two pixels (**c**) a schematic of the p + n Si pixel (strips used) sensors with a thin metal layer on top of the oxide surface

Table 7.1 Specification of AGIPD sensor with an aim for 1 GGy radiation tolerance

	Parameter	Value	Criteria	Comments
1.	Thickness	0.3–1 mm (0.5 mm)	X-ray absorption length; radiation exposure electronics	0.5 mm for first sensors
2.	Pixel size	0.2·0.2 mm^2		Fit to readout electronics
3.	Crystal type	p^+n	Radiation hardness, plasma effects polarity of pulse	Standard—bigger choice of manufactures
4.	Depletion voltage	200 V		
5.	Operating voltage	<1000 V	Reduced plasma effect minimize accumulation layer	Design challenge
6.	Dead edges	<0.5 mm	Science, tiling	Design challenge
7.	Passivation	SiO$_2$/Si$_3$N$_4$	Vacuum operation, environmental effects, insulation, radiation hardness	Depends on manufacturer
8.	Biasing scheme	DC	Electronics (leakage current)	
9.	Interpixel C	<0.5 pF	Noise ad cross talk	
10.	Max. dark current/ pixel	<1 nA	Noise, pedestal stability	Stable within <30%
11.	Max. total current	100 µA	Heat, safety, machanics	
12.	Flatness	<25 µm	Bump bonding	

7.4 Simulation Results and Discussions

The two-dimensional simulations of Si pixel sensors (Fig. 7.1) were performed in order to investigate the influence of operating voltage, metal overhang width on the WEAL (width of electron accumulation layer), electrical breakdown voltages V$_{BD}$, dark surface generation-recombination current (I$_{ox}$), and interpixel capacitance (C$_{int}$) of non-irradiated (0 Gy) and irradiated (5 MGy) Si sensor.

At the European Free-Electron-Laser XFEL the sensors will be exposed to surface doses of up to 1 GGy of 12 keV X-rays. For this X-ray energy displacement damage effects can be excluded. I/V-, C/V- and TDRC-(Thermally Dielectric Relaxation Current)-measurements on gate controlled diodes and CMOS capacitors fabricated by CiS on 2 kΩcm n-type float zone (FZ) Si have shown that oxide charge density, interface trap density and surface generation velocity saturate, and even decrease beyond a dose of several MGy [4]. For simulation of non-irradiated sensor (0GY), we use only an extracted value of an oxide charge density of $N^{fix}_{ox} = 2.1 \times 10^{10}$ cm^{-2} from the comparison of the simulation and measurement of 1/C^2 versus of operating voltage. Based on the results at 5 MGy we use for the simulations an oxide charge density of $N^{fix}_{ox} = 2 \times 10^{12}$ cm^{-2} and two Gaussian shaped interface trap level distributions with the parameters given in Table 7.2.

Table 7.2 Microscopic parameters of the interface trap levels derived from measurements of CMOS capacitors irradiated to 5 MGy used in the TCAD simulation

Interface trap type	Energy level [eV] E_c-E_{it}	σ_{it} [eV]	D_{it} [cm^{-2} eV^{-1})	σ_{eff} [cm^2]
Acceptor	E_c-0.354	0.1016	5.861×10^{12}	1×10^{-16}
Acceptor	E_c-0.636	0.045	6.728×10^{12}	4×10^{-15}

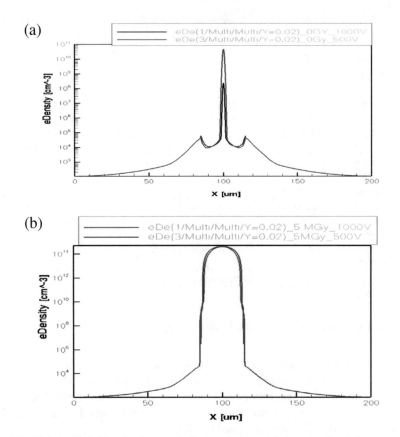

Fig. 7.2 (Color online) (**a**) e-density versus distance (x in μm) for non-irradiated Si sensor as a function of operating voltage. (**b**) e-density versus distance (x in μm) for 5 MGy irradiated Si sensor as a function of operating voltage

The first simulation result for no metal-overhang width is shown in Fig. 7.2, which shows the plot of e-density versus distance along in x-direction for two operating voltages (500 V, and 1000 V). The results indicate that there is an appreciable change in the e-density and also in the WEAL in 0Gy sensor with increasing voltage from 500 V to 1000 V whereas no change has been observed in the WEAL at 5 MGy irradiated sensor with increasing voltage from 500 V to 1000 V. This is due to the fact that undepleted region at the Si-SiO$_2$ interface at 5 MGy. Consequently, for a given metal-overhang width the WEAL decreases in

Si-bulk with increasing operating voltage, resulting in higher CCE (as expected []) because metal-overhang act as a reverse bias of MOS (Metal-Oxide-Semiconductor) structure and the tendency is to deplete the surface region at the Si-SiO$_2$ interface.

7.5 Technological Implication

The following implications on the physical and geometrical parameters such as junction depth, thickness of oxide passivation, operating voltage, WEAL, and dead edges width for the surface radiation hard design of the AGIPD sensor are suggested.

7.5.1 Influence of Junction Depth (X$_J$)

If the junction depth is increased, the electric flux per unit area at the curved portion of the junction decreases and therefore the field crowding reduces, which in turn increases the avalanche breakdown voltage (V$_{BD}$). Junction depth is not an important design parameter for the surface damaged sensors. It is just choice of vendors technology i.e. 1 μm for HPK technology (for reduced dead space). This is because of the as expected result:

- no effect on WEAL (width/depth of electron accumulation layer).
- increases of avalanche breakdown voltage. The effect will be different for different oxide+nitride thickness and metal-overhang width.

It has been already shown that for two junction depths in no guard ring based pixel sensor design, no V$_{BD}$ is observed up to 1000 V because of the reduce peak of an electric field in the presence of interface traps for different gaps. It means that we can choose the 1 μm junction depth (shallow junction devices).

7.5.2 Thickness of Oxide Passivation (t$_{ox}$)

When ionizing radiations penetrates the Si sensors, it will create the e-h pairs in the oxide. The electron will pass through the circuit due to its high mobility but the holes will be trapped in the oxide near to Si-SiO$_2$ interface.

The oxide trapping is depending upon the oxide thickness (tox [1.4] and this will also change the threshold voltage of the device. In the present work, similar oxide +nitride thickness is used in the TCAD simulation as per the thickness of the measured CMOS capacitors in order to understand the effect of microscopic defects on the valuable parameters as shown above. The technological suggestion is to use minimum oxide thickness of an order of less than 200 nm in order to reduce the threshold voltage up to ~1.5 V instead of 15 V in irradiated sensor (10 Mrad).

7.5.3   Operating Voltage

It should be noted that for the final design of the radiation hard p^+n pixel sensors: the operating voltage should be 1000 V or around 500 V in order to avoid high voltage sparking (HV) sparking at cut edge (sufficient to avoid e-h plasma effects (charge explosion effects) at hard XFEL for the XFEL spot size! [5]).

7.5.4   WEAL (Width of Electron Accumulation Layer)

The aim is to reduce the width of the WEAL in the order of few microns for no hole charge loss in the EAL. That can be only possible in the lowest gap with large overhang. The aim is to understand how to reduce WEAL at 1000 V operating voltage (a one of the major design challenge)?. It has been found that the WEAL changes with increase of W_{MO}. But for shorter W_{MO} there is no appreciable change in the WEAL whereas for the larger $W_{MO} = 12.5$ μm, a small decrease in the WEAL observed (see Fig. 7.3a). This is due to that fact of the metal overhang width on oxide act as gate boundary conditions and that is changing the surface potential on the oxide surface and thus depletes the Si-SiO$_2$ surface area and the main consequence is the reduction of the WEAL.

It is known from our previous chapter the generation-recombination surface current (I_{ox}) increases with the operating voltage and this will originate from the depleted Si-SiO$_2$ surface area. It can be seen in Fig. 7.3b that the shape of the current-voltage characteristics is strongly influenced by the W_{MO}. For no metal overhang width, I_{ox} increases with the operating voltage. With increasing W_{MO} depleted Si-SiO$_2$ surface area also increases and thus the outcome is the increases of the I_{ox} and for higher operating voltage, a flattening in the current-voltage characteristics is observed for the larger W_{MO}.

For the major reduction in the WEAL, we have proposed the two radiation hard designs (Rad-hard method I—Fig. 7.1b and Rad-hard method II—Fig. 7.1c). In both Si sensor designs, a larger overhang of 12.5 μm is used for the estimation of the WEAL and the expected value of the WEAL can be also extrapolated for the lower W_{MO} at different operating voltages (500 V to 1000 V).

For the radiation hard method I, the devices can be built with the two alternatives approaches. The fabrication process can be simulated if the fabrication parameters (implantation energies, doses, diffusion temperature, and during oxidation) are known in detail. If this is not possible, the devices can be built directly though definitions of regions and their properties (doping, implant size, depths, profiles, p-stop dose etc.). Second approach has been used in this work. Several p-stop dose have been used to identify the WEAL if it is compensates the N_{ox}^{fix}. Here, we have

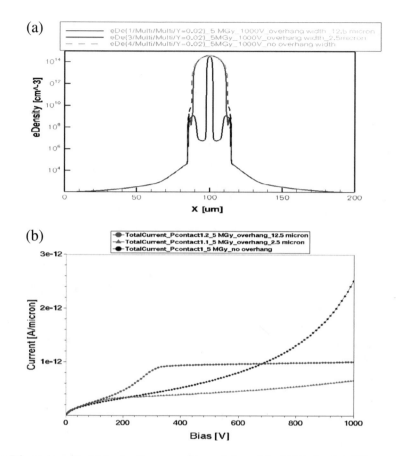

Fig. 7.3 (Color online) (**a**) e-density versus distance (x in μm) for 5 MGy irradiated Si sensor as a function of metal overhang width. (**b**) Current-voltage characteristics of 5 MGy irradiated Si sensor as a function of as a function of metal overhang width

shown the best results for the 10 μm width of p-stop (dose $= 1.125 \times 10^{15}$ cm^{-2}) for the present Si pixel sensor design simulation.

It can be seen in Fig. 7.4a that for rad-hard method I, p-stop compensates the EAL density and with the increasing voltage, peak concentration of electron are also decreases whereas for rad-hard method II, WEAL also reduces with increasing operating voltage (Fig. 7.4b).

It has been found that lower I_{ox} current is obtained for the rad-hard method I than II (see Fig. 7.5) and increases with the increase of the operating voltage and flattened at very high more than 800 V but for both methods, I_{ox} are within the sensor design specifications (Table 7.1).

Now, it is important to see the sensor design influence on the inter pixel capacitance and cross-talk which is a creation of noise in the electronics. We have performed the small signal ac analysis at 1 MHz frequency in order to find the

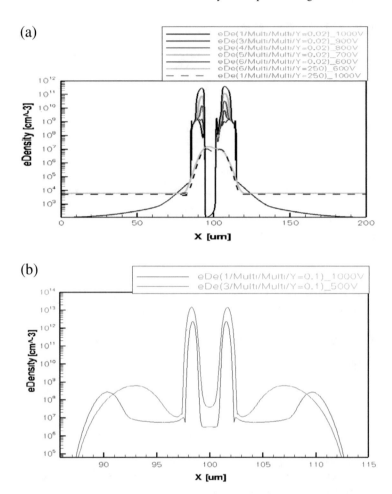

Fig. 7.4 (Color online) (**a**) e-density versus distance (x in μm) for 5 MGy irradiated Si sensor as a function of operating voltage—Rad-hard method I. (**b**) e-density versus distance (x in μm) for 5 MGy irradiated Si sensor as a function of operating voltage—Rad-hard method II

capacitances and conductance's matrix at different voltages between different adjacent electrodes. Table 7.3 shows the interpixel capacitances value for the both radiation hard methods for non-irradiated (0 Gy) and irradiated sensors (5 MGy) at 60 min (80 °C).

It is observed that similar and lower interpixel capacitance obtained in non-irradiated and irradiated Si pixel sensors for rad-hard method II than rad-hard method I. This is one of a challenging design for the low noise readout electronics with low cross-talk for high resolution X-ray imaging detectors. In future, a test structure will developed in order to compare the simulation result with experimental data.

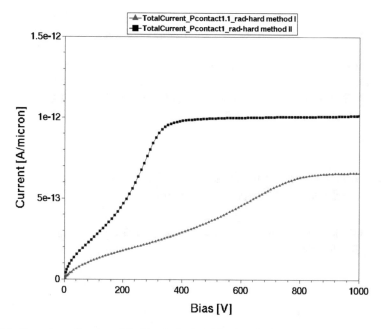

Fig. 7.5 Current as a function of the bias for both rad-hard method I and II

Table 7.3 Shows the interpixel capacitances values for the both methods for different irradiation doses (0GY and 5 MGy, 60 min 80 °C)

S.No.	Normal		Rad-hard method I		Rad-hard method II	
Irradiation doses	0 Gy	5 MGy	0 Gy	5 MGy	0 Gy	5 MGy
Interpixel capacitances ($4C1 + 4C_{diag}$)(fF)	120	129	288	267	101	101
$C_{metal-implant}$ $(4C_m\text{-}p^+)$					38.32	39.12
Back-plane Capacitances (fF)	8.36	8.36	8.36	8.36	8.12	8.12

It is clear that for the rad-hard method II, I_{ox}, capacitance (within specification) is independent of the operating voltage (from 500 V to 1000 V) for a large metal overhang width of 12.5 μm.

The following technological things are suggested to reduce the WEAL (concentration of e^-) in surface damaged Si sensors:

- Large metal overhang width
- High bias voltage (500V to 1000 V!)
- Rad-hard method II (a thin metal on top of electric field minimum (Emin) valley) with low cross-talk.

For 1000 V operating voltage, the following values should be noted, WEAL = 1.05 μm, depth from X = 98.5 μm, 0.115 μm/pixel for 12.5 μm metal overhang width.

For the radiation hard method II, an equivalent 3-D total pixel capacitance without error implications in 2-D and 3-D simulation result, total capacitance is $C_t = 109.12$ fF, and $I_{dark-current} = 0.2$ nA (depleted area = 2.15 μm^2), this will change with the different operating voltage.

7.5.5   Dead Edges: Simulation for 0.5 mm Dead Edges and 1000 V Guard Ring Designs

The TCAD simulation for 0.5 mm dead edges and 1000 V guard ring designs are still ongoing for that kind of design.

7.6   Conclusion

The present work is an effort aimed to optimize the radiation hard sensor design for the AGIPD detector at the European XFEL which will give the reduced WEAL up to a few μm in order to avoid hole change loss and low interpixel capacitance at 5 MGy dose. Based on the 2-D computer simulation result, a *layout* of a radiation hard sensor design-II with optimized performance is proposed.

Acknowledgement The author would like to thank the XFEL company for support and also would like to thank to the peoples involved in the development of AGPID for XFEL experiment from DESY (Deutsches Elektronen Synchrotron), PSI (Paul Scherer institute), Switzerland and University of Bonn, Germany for constant interest and support. This work was profited from the infrastructure grant of the Helmholtz Alliance "Physics at the Terascale".

References

1. The European X-Ray Laser Project *XFEL*. http://xfel.desy.de/
2. AGIPD (Adaptive Gain Integrating Pixel Detector). http://hasylab.desy.de/instrumentation/detectors/projects/agipd/
3. Synopsys Inc., TCAD software. http://www.synopsys.com/Tools/TCAD/DeviceSimulation/
4. Fretwurst, E., Januschek, F., Klanner, R., Perrey, H., Pintilie, I., Renn, F.: Study of the radiation hardness of silicon sensors for the XFEL, poster presented at the Nucl. Sci. Symposium IEEE, Dresden 2008, Germany, Conf. record N30–400
5. Becker, J., Eckstein, D., Klanner, R., Steinbrück, G., on behalf of the AGIPD Consortium: Impact of plasma effects on the performance of silicon sensors at an X-ray FEL. Nucl. Instr. Methods Phys. Res. A. **615**(2), 230–236 (2010)

6. Richter, R.H., Andricek, L., Gebhart, T., Hauff, D., Kemmer, J., Lutz, G., Weiß, R., Rolf, A.: Strip detector design for ATLAS and HERA-B using two-dimensional device simulation. Nucl. Instr. Methods Phys. Res. A. **377**, 412–421 (1996)
7. http://hasylab.desy.de/instrumentation/detectors/projects/agipd/presentations/e97045/ASrivastavaDevelopmentofRadiationHardSiPixel.pdf
8. Ma, T.P.: Generation and transformation of interface traps in MOS structures. Microelectron. Eng. **22**, 197–200 (1993)
9. Srivastava, A.K., et al.: Numerical modelling of Si sensors for HEP experiments and XFEL (POS RD09) 19 (2010)
10. Wunstorf, R.: Ph.D. thesis, University of Hamburg, DESY FHIK-92-01, October 1992
11. Longoni, A., Sampietro, M., Struder, L.: Instabilities of the behaviours of high resistivity Si detectors due to the presence of oxide charges. Nucl. Instr. Methods Phys. Res. A. **228**(2), 35–43 (1990)

Chapter 8
Capacitances in P⁺N Silicon Pixel Sensors Using 3-D TCAD Simulation Approach

8.1 Introduction

At XFEL experiment, sensors should have high voltage stability up to 500 V to avoid charge explosion effect [1] with some safety margin and should have low interpixel capacitance up to 0.5 pF [2]. In our previous simulation approach, we have proposed the optimum gap for the low interpixel capacitance [3]. Here we have used the optimum gap spacing between the adjacent p⁺ pixels for the capacitance calculations. The cross-talk effect is not taken into considerations.

The capacitance calculations using ISE TCAD software, HSPICE [4], and analytical calculations using 3D Laplace equation [5] is already performed for the rectangular and square type of pixels.

Here we have also shown the capacitance calculations for 3×3 sensor pixel array and the results are presented.

8.2 Device Design and Simulation Technique

The p⁺n Si pixel sensor is made on 500 µm thick n-type high resistivity Si material (3–4 kΩ-cm) which is an equivalent to substrate doping concentration of 1×10^{12} cm⁻³. The p⁺ impurity profile is approximated by assuming a Gaussian profile with a peak concentration of 5×10^{19}/cm³ at surface and at junction 1×10^{15}/cm³. The same doping profile is assumed in all p⁺ pixel regions. The backside is implanted with n⁺ of thickness 1 µm and then metalized with Aluminium (Al) of thickness 1 µm to take ohmic contact. It is assumed that lateral diffusion depth at the curvature of the p-region is equal to 0.8 times the vertical junction depth. p⁺ pixel implants are grounded through an ohmic contact (not shown in the figure) and are DC-coupled to the Al metal. The Fig. 8.1 is showing the top view of the 5×5 sensor pixel array and the floating guard rings surrounds the pixels but 3×3 portions of

© Springer Nature Switzerland AG 2019
A. K. Srivastava, *Si Detectors and Characterization for HEP and Photon Science Experiment*, https://doi.org/10.1007/978-3-030-19531-1_8

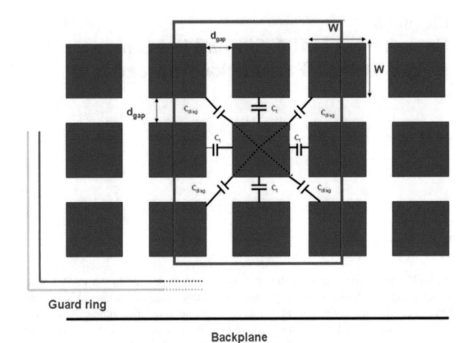

Fig. 8.1 Layout of 5 × 5 sensor pixel array. The simulated portion is marked by red colour

$400 \times 680 \times 500 \, \mu m^3$ is simulated here because the effect of second neighbors is not important in the square pixel geometry. In the Fig. 8.1, d_{gap} is showing the distance between the pixels, W is the width of the pixel. The rest symbols used in the Fig. 8.1 are shown below: The capacitances are as follows:

C_0—Back-plane capacitance,
C_1—Capacitance between adjacent pixel,
and C_{diag}—Diagonal capacitance.

The sensor is simulated using Synopsys TCAD 2010.03 [6] commercial device simulation program.

The SRH (Shockley Read-Hall Recombination) recombination, Auger recombination, Impact ionization, doping dependent mobility, high field saturation physical models are used to initiate the device simulation program. The default Dirichlet and Neumann Boundary condition is applied at contacted and non-contacted edge of the sensor pixel array. The small signal AC analysis is performed for the capacitance calculations at 1 MHZ frequency to measure the real and imaginary part of the admittance at different voltages in between the different resistive electrodes. For the comparison with published analytical calculations, an ideal condition of the sensors is used.

8.3 Physics of the Pixel Capacitances

A crucial factor in evaluating the noise contribution of DC coupled p^+n Si pixels sensor is the sum of the value of the direct capacitances between adjacent and diagonal electrodes called interpixel capacitance (C_{int}). The total capacitive load for sensor pixel array can be approximated by $C_0 + 4 C_1 + 4 C_{diag}$. The effect of other long placed pixels on the center pixel can be neglected.

The relation of equivalent noise charge (ENC) with total detector capacitance (C_D) is by $ENC^2 \propto 24kT(C_D + C_{FET})^2/(3g_m)$, where C_D consists of two terms backplane capacitance and interpixel capacitance, C_{FET} is the capacitance associated with the input FET of the preamplifier, k is the Boltzmann constant, T is the absolute temperature and g_m is the transconductance of the input FET. The significance of studying C_{int} *is further attributed to its creation of cross talk* [7, 8]. The cross-talk can be calculated by $C_1/C_{bump.bonded} + C_{pre-amp} + C_{total}$. The cross talk can reduced by increasing the total capacitance like an addition of some metal on ground routing between the pixels. Thus, for the above reasons it is important to investigate the characteristics of the Si pixel sensor in terms of the interpixel capacitance. The present work is an attempt in this direction. In this paper, 2-D numerical device simulation using Synopsy TCAD 2010.03 has been exploited in order to evaluate the mutual capacitance between two facing strips. Through small signal AC-analysis, the admittance matrix of the network shown in Fig. 8.1 and it can be solved at an arbitrary bias point, from which mutual capacitance and conductance between the facing and diagonal pixels can be estimated. In AC coupled Si pixel sensor, the contribution to interstrip capacitance (C_{int}) between adjacent and diagonal strips mainly comes from four components: (1) The capacitance between metal of ith and jth strips ($C_{Mi - Mj}$), (2) the capacitance between the implanted strips ($C_{Ii - Ij}$), (3) the capacitances between a metal and adjacent strip's implant ($C_{Mi - Ij}$, $C_{Mj - Ii}$) and (4) the coupling capacitances between a metal strip and implanted strip ($C_{Mi - Ii}$, $C_{Mj - Ii}$). However, it should be noted that the coupling capacitances $C_{Mi - Ii}$ and $C_{Mj - Ij}$ are usually much larger than the remaining parasitic components in order to avoid spreading of the signals on the pixels, so that we can consider their impedance to vanish in the high frequency range of interest. Thus, to first order approximation, we can assume that the total interpixel capacitance between two facing and diagonal pixels is given by $C_{int} = C_{Mi - Mj} + C_{Ii - Ij} + C_{Mi - Ij}$, *where*, $C_{Mi - Ij}$ includes the contribution of both the capacitances between $M_i - I_j$ and $M_j - I_i$. The value of the capacitance between the read out electrodes $C_{Mi - M}$ is usually smaller than other two capacitances because of the air dielectric capacitances. The shape of the CV (capacitance-voltage) curve for the different capacitances between the electrodes is depends upon the several parameters like width by pitch ratio, metal over hand width, surface irradiation doses, bulk related deep trap effects after hadronic irradiations etc. At XFEL experiment, the surface damage effects (N_{ox}, D_{it}) in silicon pixel sensors is dominant therefore two lumped, three lumped model or several can be drawn in order to understand the shape the C/V curve. Here, we have proposed the two lumped models for two facing pixels (see Fig. 8.2).

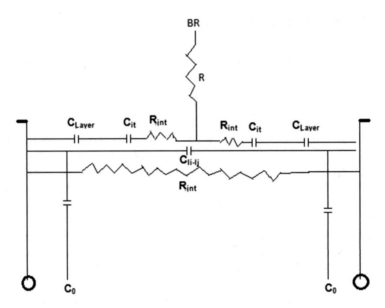

Fig. 8.2 Two lumped model for the C/V curve description of two adjacent strip/pixel sensor

In the presence of surface charge effects, an accumulation layer of fixed oxide positive charge and interface trap at Si-SiO_2 interface will be developed. In Fig. 8.2, it can be see that the net list of RC network of physical parameters in two strip-pixel subset of sensors. In Fig. 8.2, C_{Layer} is the capacitance between the pixel and accumulation layer of free electrons, R is the resistance of the accumulation layer and C_{it}, R_{it} is the capacitance and resistance of the interface trap.

The interpixel resistance is represented here by symbol R_{int}. The left and right part of the Fig. 8.2 can be expressed as C′ and C″ and the C′ is serial combination of C_{int}, C_{it}, and C_{Layer} and similarly the second right part of the Fig. 8.2. The total parallel capacitance is the sum of the serial combination of C′, and C″ and the conductance is the sum of all inverse resistances.

8.4   Results

In this work, sensors pixel array is simulated using Synopsys TCAD for the capacitance calculations. The electrostatic potential distribution, E-field, and e-concentration is shown in order to see the depletion behaviour and magnitude of the E-field inside the sensor (see Fig. 8.3).

The change of the interpixel capacitance in between the different adjacent electrodes and diagonal with applied bias voltage is shown in the Fig. 8.4. The initial decrease of the capacitance with voltage is due to the capacitance contribution from the back-plane of the sensor and for further an increase of the voltage the

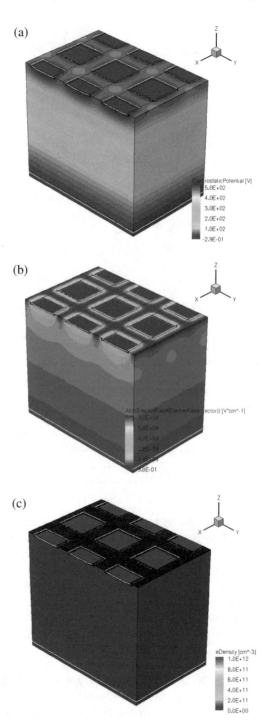

Fig. 8.3 (a) Electrostatic potential (b) E-field (c) e-concentration inside sensor pixel array at 500 V bias voltage and 1 MHz frequency

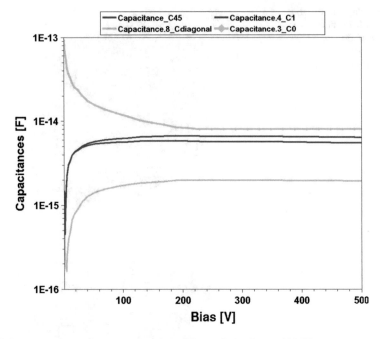

Fig. 8.4 Interpixel capacitance as a function of the applied voltage at 1 MHz

Table 8.1 Comparison of 3-D simulated capacitances with analytical expressions for 80 μm p⁺ pixel gap

Capacitances	Analytical calculation [4] in fF	Simulation [S] in fF	Error [%]
C_0	7.68	8	5
C_1	5.099	5.48	22
C_{diag}	1.355	1.94	34

capacitance increases because of the contribution of the capacitance from the backplane into the direct interpixel capacitance because of the small ac signal is coupled to the back-plane (see Fig. 8.2 for explanation). Table 8.1 shows the simulated back-plane and first neighbour's capacitance, and diagonal capacitance. In an irradiated sensor, frequency effects can be easily explained from the model shown in Fig. 8.2 with the C_{Layer}, and C_{it} parameter.

It has been found that the there is an agreement with 5% with back-plane capacitance whereas in the capacitance for the first neighbours is within 22% and 34% in diagonal capacitance (Table 8.2).

It has been found that the simulated back-plane capacitance using 2-D (normalized for 200 × 200 μm²) and 3-D Synopsys TCAD simulation is in almost 100% good agreement with the analytical expressions [3] where from [1], the agreement is in within 5%. The agreement within 25% for 2-D and 3-D simulated total capacitive

Table 8.2 Comparison of 2-D and 3-D simulated back-plane and interpixel capacitances with analytical expressions for 80 μm p$^+$ pixel gap

Capacitances	Analytical calculation in fF		C_{Sim}-Synopsys (2D) in fF	C_{Sim}-Synopsys (3D) in fF	Implications
Back-plane capacitance (C_0)	$C_{pixel} = C_{strip}*$Length Cstrip is in good agreement with theoretical calculations [9]	[5]	8.28	8	~100% good agreement in 2-D and 3-D simulation result and [3] Analytical calculation result [1] and with [3] and 3-D simulation, agreement within 4%
	8.3	7.68			
Total capacitive load (without approx 100 fF from bump bonding +pre-amp load)	33.438 [5]		53	39.9	Agreement between 2-D and 3-D simulations within 25% and analytical calculations within 16% – Results are consistent with [2]

load and within 16% with analytical expressions given in [1]. The results are consistent with previous published work [2].

8.5   Conclusion

The first AGIPD pixel sensor prototype fabricated by Hamamtschu (HPK), JAPAN will consist of p$^+$ pixels in the n-bulk because of the successful operation up to 500 V without any avalanche breakdown and low fabrication cost, easy fabrication technique and expertise has expected no problem of sparking at the edge of the sensors due to the guard-ring biasing scheme. This optimized technology with 80 μm optimum gap ensures low total capacitive load up to 140 fF including contribution from bump bond (can be estimated) and charge sensitive pre amp load at the readout of ASIC electronics and the observed results are consistent with the previous published work.

   In our previous work we have already optimized the sensor design using 2-D simulation approach with sufficient radiation hard hardness up to 5 MGy X-ray dose (annealed 60 min 80 °C). Therefore from the present analysis, total capacitive load of 250 fF is estimated in the breakdown protection field plated sensor array structure irradiated by 5 MGy X-ray dose (annealed 60 min 80 °C).

Acknowledgement The author would like to thank the XFEL company for support and also would like to thank to the peoples involved in the development of AGPID for XFEL experiment from DESY (Deutsches Elektronen Synchrotron), PSI (Paul Scherer institute), Switzerland and University of Bonn, Germany for constant interest and support. This work was profited from the infrastructure grant of the Helmholtz Alliance "Physics at the Terascale".

References

1. Becker, J., Eckstein, D., Klanner, R., Steinbrück, G., on behalf of the AGIPD Consortium: Impact of plasma effects on the performance of silicon sensors at an X-ray FEL. Nucl. Instr. Methods Phys. Res. A. **615**(2), 230–236 (2010)
2. Srivastava, A.K., Eckstein, D., Fretwurst, E., Klanner, R., Steinbrück, G.: Numerical modelling of Si sensors for HEP experiments and XFEL. POS (RD09) 019, 2009
3. Srivastava, A.K., et al.: Development of radiation hard Si pixel sensor for the 4th generation photon science experiment at XFEL, (Internal note)
4. Gorfine, G., Hoeferkamp, M., Santistevan, G., Seidel, S.: Capacitance of silicon pixels. Nucl. Instr. Methods Phys. Res. A. **460**, 336–351 (2001)
5. Cerdeira, A., Estrada, M.: Analytical expressions for the calculation of pixel detector capacitances. IEEE Trans on Nuclear Science. **44**(1), 63–66 (1997)
6. Synopsys Inc., TCAD software. http://www.synopsys.com/Tools/TCAD/DeviceSimulation
7. Chatterji, S., Bhardwaj, A., Ranjan, K., Namrata, Srivastava, A.K., Shivpuri, R.K.: Analysis of interstrip capacitance of Si microstrip detector using simulation approach. Solid State Electron. **47**, 1491–1499 (2003)
8. Rohe, T., Hügging, F., Lutz, G., Richter, R.H., Wunstorf, R.: Sensor design for the ATLAS-pixel upgrade. Nucl. Instr. Methods Phys. Res. A. **409**, 224–228 (1998)
9. Dell' Orso, R.: Recent results for the CMS tracker silicon detectors. CMS Conference Report CMS CR 20001/003

Chapter 9
Characterization of Si Detectors

9.1 Introduction

The LHC (LARGE HADRON COLLIDER) at CERN, Geneva is one of the prestigious High-Energy physics (HEP) collider experiment. The LHC (pp Collider, 14 TeV, 25 ns bunch spacing) is foreseen to be upgraded to HL-LHC, where the luminosity increases of up to ten times i.e., 10^{35} cm^{-2} s^{-1} [1, 2]. At HL-LHC up to 400 interactions per bunch crossing are expected. This causes a major increase in track density, requiring for intermediate and larger radii smaller detection elements with higher granularity than the present the CMS silicon tracker. The radiation damage effects in the Si sensors at HL-LHC will be more challenging to cope with such hostile radiation environment therefore the Compact Muon Solenoid (CMS) experiments will require a new CMS tracking detectors in the phase 2 upgrade (2026) of the HL-LHC.

The bulk radiation damage affects the Si sensors performance in terms of increases of leakage current due to increases of generation recombination centers, increases of full depletion voltage (V_{FD}) due to change of effective doping concentration, and degradation of charge collection efficiency (CCE) due to charge carries trapping [3–7]. The Double Junction (DJ) effect has been observed in irradiated Si sensors, which will give Double Peak (DP) electric field due to the occupation of deep traps. This phenomenon is also referred to type inversion or space charge sign inversion (SCSI); Si sensors will undergo type inversion thus for higher CCE, sensors should be over depleted for the high electric field over the entire sensor volume. The leakage current can be reduced by operating temperature below of $-30\,^{\circ}$C [3–7]. Therefore sensor design should have capability to withstand the high voltage operation for the successful operation of the experiments. For the improve-

© Springer Nature Switzerland AG 2019
A. K. Srivastava, *Si Detectors and Characterization for HEP and Photon Science Experiment*, https://doi.org/10.1007/978-3-030-19531-1_9

ments in the avalanche breakdown voltage up to more than 1000 V, a lot of techniques have been known like junction termination techniques, multiple field limiting guard rings structures at cut edge, final passivation layer technology dependence on the breakdown voltage etc. [A.K. Srivastava].

The surface damage introduces surface charges (oxide charges in SiO_2 and interface charges at Si-SiO_2 interface) that will degrade the macroscopic performance of Si sensors: high field at critical corner that will lead to avalanche breakdown, increases of depletion voltage and surface current, change of interstrip capacitance and interstrip resistance [8].

The TCAD device and process simulation [9] is an inexpensive way to develop optimized rad-hard design before fabrication of the detectors as per the radiation requirement of the HL-LHC collider experiments [A.K. Srivastava]. A lot of scientists are involved in the designing of detectors using TCAD simulator and finally, the optimized process & device physical parameters sent to foundry for the fabrication of the Si detectors.

The characterizations of the non-irradiated and irradiated detectors are one of the most important main concerns of the detector physicist to check the performance of the detectors in the harsh radiation environment of the HL-LHC.

9.2 Review of Status of Research and Development in the Subject

9.2.1 International Status

In the framework of the CERN RD50 collaboration, a lot of progress has been made in recent years in order to improve the radiation hardness of Si sensors for very high luminosity colliders since 2002 and still work ongoing to choice the intrinsic detector type for the upgrade of the particle tracking systems of all the LHC experiments in order to achieve good position resolution, acceptable charge collection (high signal to noise ratio) with low material budget technology [4]. The material/defect engineering is one of R & D approach where several microscopic defects identified which has energy levels in the band gap of Si and have macroscopic effects on the Si sensor performance [5] and device engineering is one of strategy to improve radiation hardness of sensors [1]. In the mixed irradiations, the use of n-magnetic Czochralski (MCz-n) Si is a material prime candidate for the SLHC to improve the radiation hardness of Si sensors [6]. Si sensor design for MCz material is still under serious investigations under the CERN RD50 collaboration and it has been already found that p-type Si sensors has improved CCE due to higher electron mobility of signal electrons than n-type and lack of type inversion effect after heavy irradiations.

In the framework of the CEC collaboration (within CMS), test structures of strixel geometry proposed and first attempt of simulation results on non-irradiated near-far strixel geometry have been reported [7]. The Si strixel sensor design with double metal layer with cross metal routing through via contact over the strip is also proposed as a new readout architecture for the large radii of a new CMS tracker for HL-LHC whose performance has to verify using simulation approach.

On the basis of a lot of R & D work which is already performed within the framework of the CERN RD50 collaboration, SMART Italian project, WODEAN, and CEC collaboration within CMS, the following things are required for the technology development of particle tracking detectors for the CMS tracker upgrade at HL-LHC.

The neutron irradiation and mixed irradiation four level deep trap model for n-type MCz Si using Synopsys TCAD is developed for the study of non-homogeneous distribution of space charges, electric field, and charge carrier trapping inside the irradiated sensors [9].

This experience is used to sensor design, build, and test the new segmented Si sensors for the big detector system and the aim is to developed technologies for the segmented Si sensors with unprecedented precision in position resolution. All aspects are viewed for the development of the radiation hard Si sensors with the goal to combine as much functionality as possible, as per their electrical and mechanical requirement for the CMS tracker upgrade.

In this context, the following things are concealed on the basis of experimental results on Si detectors especially for the development of n-MCz Si and p-MCz Si sensor technology for the high luminosity collider experiments and that is taken into account for better understanding of the current picture on Si sensor design:

- Trapping should be reduced or control for sufficient high CCE.
- The knowledge of physical structure of cluster defects are necessary for the improvement in radiation hardening of sensors.
- Thin p-in-n MCz Si sensors (strips and strixel) with n-side readout at HL-LHC fluence will be a future promising option in a future tracker of CMS.
- -n-in-p MCz Si sensors could be also choice for outer layer [10], probably even for inner radii with high bias voltage at HL-LHC and it should be used and test for further improvement in sensor design.
- no type inversion.
- design risk (optimal radiation hard design for breakdown voltage >1000 V).
- CCE do not anneal.

- at high bias voltage (may be breakdown will occur). This is because of the high reverse annealing growth.
- Higher CCE obtained for thin sensors than standard thickness of 250–300 μm (reduced multiple scattering at HL-SLHC) due to avalanche multiplication thanks to their low material budget.
- thin n-in-p pixels could be replaced the n in n pixels due to simplicity of design and higher charge collection of 15ke @700 V in the test beam with a track reconstruction efficiency of 99% reported [11] and high voltage sparking at cut edge can be reduced for e.g. by a parylene coating of the edge region after bump-bonding! so alternative isolation method can be explored.
- Reduced inactive dead edge width <500 μm reduces the geometric inefficiency of the Si sensors than usual 1 mm wide dead region.

9.3   Methodology for Designing of Detector

The rad-hard detector designs (centre & cut edge of the device) with Multiguardring structures for the HL-LHC will be produced on p on n-type, or n-or p-type high resistivity MCz Si <111> or <100> Si materials. The chosen process and device parameters will be used for the designing & optimization of the detector design using Synopsys TCAD device simulation. The idea is to optimise the rad-hard detector design (p on n-type, or n-or p-type high resistivity MCz Si—Ist and II metal) by changing the process and device parameters like oxide thickness, junction depth, fixed oxide charge density N_{ox}, interface trap density D_{it}, overhang width, gap between the strips/pixels, p-spray dose, guarding spacing, guard ring width etc. It is known that computer

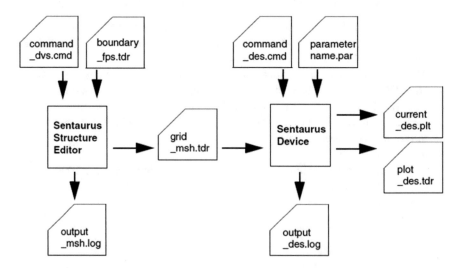

Fig. 9.1 Typical tool flow with Synopsys TCAD device simulation using Sentaurus device

simulation using commercial simulation program like Synopsys TCAD (Sprocess and Sdevice simulation) offers a potentially powerful tool for designing new semiconductor devices or processes. The use of process and device modeling for computer simulation is an inexpensive way to design and simulate modern device structures that are so complex and the right way to solved the design problem and improve the device performance and here we have used as one of the a comprehensive research program of radiation damage in silicon sensors. Figure 9.1 shows the typical tool flow with Synopsys TCAD commercial device simulation program using Sentaurus device [8]. The simulator is already rigorously calibrated with experimental data [8].

The impression is to simulate the various sensor designs and its test structures (non-irradiated, $N_{ox} = 2 \times 10^{10}$ cm^{-2} ~ <100> and $N_{ox} = 3$–4×10^{11} cm^{-2} ~ <111> orientation and irradiated by bulk damage (N_{ox} = saturate up to 1×10^{12} cm^{-2} or surface damage $N_{ox} + D_{it}$ effect) to look into the sensor for potential distribution, electric field distribution, e-density, voltage handling capability, avalanche breakdown voltage, dark current, coupling capacitance, oxide capacitance, high frequency capacitance, surface current, flat band voltage, capacitance between adjacent electrodes, cross-talk, interstrip capacitance, interstrip resistance, charge collection efficiency (alpha particle). The analytical calculations will be also performed for the macroscopic parameters just for cross-check.

The detailed modelling of radiation damage can be performed by our four level neutron irradiated MCz Si model which is already proposed in [8], other trap models for n/p FZ that are developed within CERN RD50 collaboration, and also mixed irradiation model for the test structure of the upgrade of a new CMS tracking detector for the Hl-LHC.

The idea is to work on segmented non-irradiated and irradiated planar detectors i.e. n on p MCz Si devices (promising sensor types as discussed above) and thin sensors because thin sensors is collecting more changes than thin sensors and also having low full depletion voltage using TCAD simulation approach (see Appendix A.1.1) for the calculation of charge collection efficiency (alpha particle as a MIPS) and parameters as discussed above.

The optimized parameters sent to vendors for fabrication, and the testing will be performed using experimental set up as given below in the Fig. 9.2:

9.4 Experimental Measurements on Si Strip Sensor

Within CERN RD50 collaboration, the following experimental methods are adopted for the characterization of Si strip detector for the high luminosity collider and can be extensively also used for the pixel detector used in the imaging experiments. Here, we shown a few of experimental measurement techniques are in Fig. 9.2a–d.

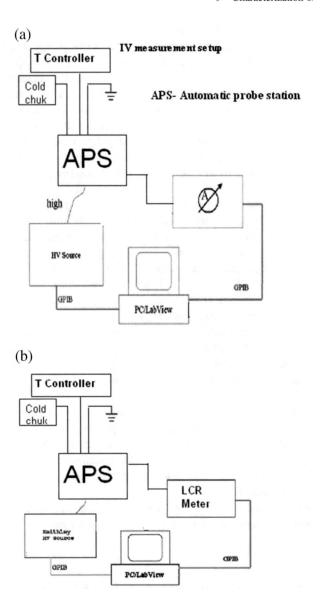

Fig. 9.2 Set for characterization of sensors (**a**) circuit for I/V (**b**) circuit for C/V (**c**) TCT circuit for CCE using laser pulse (**d**) set up for TSC measurement for microscopic effects in irradiated sensors. APS (automatic probe station) has probe needles and that will be connected with strip/backplane or bias ring of sensor to apply bias. The placement of the needles requires a good microscope and long focus

(c) Equipment to make the TCT measurements:

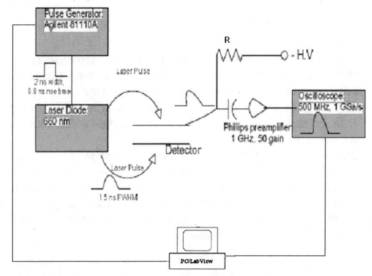

(d)

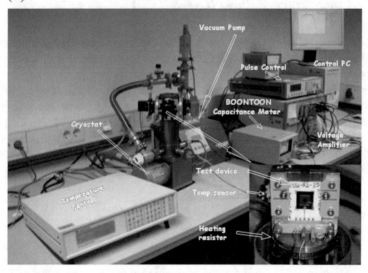

Fig. 9.2 (continued)

9.5   Brief Description of the Concept of the Set-Up

9.5.1   Current-Voltage (I-V Set Up)

For I-V measurement of the silicon sensors, the dedicated automatic probe station (APS) will be used and I will be used the probe needles to make connection of the sensor placed on cold chuck to apply a positive HV source on backplane of the sensor and p^+ implant strip is on ground potential thought he lab view program installed in the personal computer. The APS set up is attached with the current source meter and temperature controller.

The idea is to measure mainly the leakage current of sensor by applying the positive bias (>full depletion voltage) on the back plane of the sensors and p^+ implant will be on ground potential and if there is multiple guard ring then it will be floating. This can be measured for several applied high bias voltage and at several cooled temperatures (irradiated sensors) to extract full depletion voltage, leakage current, current related damage constant.

9.5.2   Capacitance-Voltage (C-V Set Up)

The same above technique is used but the LCR capacitance source meter will used to measure the capacitance between the electrodes of the test sensor at different applied frequency.

The idea is to measure mainly the full depletion voltage of sensor by applying the positive bias (> full depletion voltage) on the back plane of the sensors and sinusoidal ac signal is on p^+ implant and if there is multiple guard ring then it will be floating. This can be measured for several applied high bias voltage and at several cooled temperatures (irradiated sensors) to extract full depletion voltage, geometrical capacitance.

9.5.3   TCT (Transient Current Technique for CCE)

Charge carriers (e and h) are generated on one side of the bias sensor that will depend upon the impinging light with a short (<1 ns) pulse or alpha particles. The signal current pulse is picked up on the electrodes due to the collection of the charge carriers. This can be measured for several applied bias voltage and at several temperatures to extract trapping time constant, drift time of charge carriers to wards the opposite polarity electrodes etc.

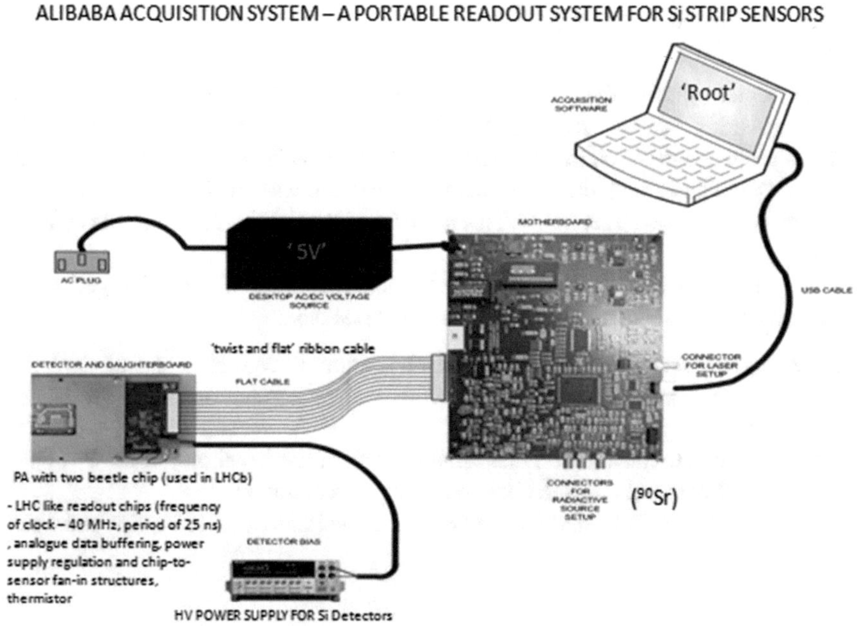

ALIBABA ACQUISITION SYSTEM – A PORTABLE READOUT SYSTEM FOR Si STRIP SENSORS

Fig. 9.3   Alibava System used for the CCE measurement

9.5.4   DLTS (Deep Level Transient Spectroscopy)

This technique can be used for sensor irradiated with low fluence $\leq 10^{12}/cm^2$. It is known that multiple deep levels will introduced in the band gap of the silicon material after irradiation of the sensors and this will be filled with charge carriers by modulating the Fermi level and measuring the C-DLTS caused by the emission of the charge carrier. The sign of the transient identifies the type of defect as hole or electron trap.

9.5.5   TSC (Thermally Stimulated Current) Measurement

This technique can be used for sensor irradiated with high fluence $\leq 10^{14}/cm^2$. Deep level defects in silicon have energy levels in the band gap. At very low temperature (5 K), defects are filled with charge carriers. With increasing temperature the charge carriers are activated and the corresponding current measured. Energy levels and defect concentrations can be extracted from the current versus temperature curve.

9.5.6 Alibava Set Up (β Set Up)

The characterization of the sensors will be carried out with the new ALIBAVA (β set up) Acquisition System (see Fig. 9.3). It is a Beetle based system to readout the signals produced in Si microstrip detectors fabricated at HPK, Japan by illuminating them with radiation sources. The system can read out, with the LHC speed electronics. The Aliabava set (β set up) will measure the collected charge versus bias voltage of strip sensor, S/N ratio etc.

The following information's can be extracted from the aforesaid experimental measurements on non-irradiated and irradiated detectors:

1. I-V—I_R (Reverse current, current related damage constant) at full depletion voltage (V_{FD}), avalanche breakdown voltage (V_{BD}), interstrip resistance (R_{int}) etc.
2. C-V—Geometrical/Junction capacitance at full depletion voltage (V_{FD}), Interstrip/Interpixel (C_{int}), Effective doping concentration.
3. Alibava set up—Collected charge versus bias voltage of detector, S/N ratio etc.

9.6 Experimental Procedures

- In irradiated detectors, room temperature annealing will start. For stop annealing, sensor will kept on freezer at −30 °C up to some time (or may be during transportation from irradiation lab to measurement laboratory) and then performed I-V and C-V measurement @RT and −30 °C (required to cool down up to this temperature for low leakage current).
- Annealing in oven up to 60 °C or 80 °C: Isothermal (T fixed, time vary), isochronal (T vary up to 400 °C, time-fixed) and then performed I-V and C-V measurement at RT and −30 °C.

The aforesaid experimental procedure is discussed here that are giving the idea about the methods to perform measurement and standard procedures to extract the experimental parameters for the analysis to get the right design for our experiments & applications.

References

1. Fretwurst, E.: Recent advancements in the development of radiation hard semiconductor detectors for S-LHC. Nucl. Instr. Methods Phys. Res. A, **A552**, 7–19 (2005). http://rd50.web. cern.ch/rd50/
2. Huntinen, M.: SLHC Electronics workshop (2004), CERN
3. Lindstrom, G.: Radiation damage in silicon detectors. Nucl. Instr. Methods Phys. Res. A. **A512**, 30–43 (2003)

4. RD50 Status Reports CERN-LHCC-2003-2004–2005-2007 [Online]. Available: http://rd50.web.cern.ch/rd50/
5. Moll, M.: PhD-thesis. Radiation Damage in Silicon Particle Detectors, University of Hamburg, Germany, 1999, DESY-THESIS-1999-40, ISSN 1435–8085
6. Kramberger, G., Cindro, V., Dolenc, I., Mandic, I., Mikuz, M., Zavrtanik, M.: Performance of silicon pad detectors after mixed irradiations with neutrons and fast charged hadrons. Nucl. Instr. Methods Phys. Res. A. **609**, 142–148 (2009)
7. Militaru, O., et al.: Simulation of electrical parameters of new design of SLHC silicon sensors for large radii. Nucl. Instr. Methods Phys. Res. A. **617**(1–3), 563–564 (2010)
8. Srivastava, A. K., Eckstein, D., Fretwurst, E., Klanner, R., Steinbrück, G.: Numerical modelling of the Sisensors for the HEP experiments and XFEL, presented at RD09 conference on 30 September–02 October 2009, POS (RD09) 019, 2009
9. Synopsys, Inc., TCAD software. http: www.synopsys.com/products/tcad/tca.html
10. Adam, W., et al.: P-type silicon strip sensors for the new CMS Tracker at HL-LHC. J. Instrum. **12**, P06018 (2017)
11. Dierlamm, A., on behalf of the CMS Tracker Collaboration: The CMS outer tracker upgrade for the HL-LHC. Nucl. Instr. Methods Phys. Res. A. **924**, 256–261 (2019)

Chapter 10
Analysis and TCAD Simulation for C/V, and G/V Electrical Characteristics of Gated Controlled Diodes for the AGIPD of the EuXFEL

10.1 Introduction

Photon Science research at the fourth generation photon science European XFEL (Eu X-ray Free Electron Laser) requires p+n Si pixel sensors at AGIPD with unprecedented performance: Doses of up to 1 GGy 12 keV X-rays and a dynamic range from 1 to 10^5 12 keV photons per sub picosecond pulse and pixel of $(200 \, \mu m)^2$ area. In this work, this is done within the AGIPD Collaboration [1–5], we present the analysis of electrical characteristics of gated diodes from the electrical characteristics of $Al/SiO_2/n$-Si (MOS) capacitors for C/V_g and G/V_g characterization. A lot of study has been made on low and high resistivity MIS capacitor in order to understand C/V_g and G/V_g behavior versus frequency [6–8].

The behavior of C/V_g and G/V_g characteristics of gated diode can be understood by MOS capacitor (heart of the gate diode). The structure can yield considerable information regarding the properties of the dielectric used, the underlying silicon and the silicon/oxide (Si/SiO_2) interface. In order to extract these properties several capacitance-voltage (C/V_g) measurement techniques have been developed. These include (i) High Frequency (HF) C/V_g measurement (ii) Low Frequency (LF) or Quasi-static C/V_g measurement (iii) Pulsed C/V_g measurement and (iv) Capacitance-time (C/-t) measurement [9–20].

This work extends for the development of radiation hard pixel sensor for the Adaptive Gain Integrating Pixel Detectors (AGIPD) of the XFEL experiment.

This paper is organized as follows: introduction is given in Sect. 10.1 and Sect. 10.2 described the design of gated diode and method. In Sect. 10.2.2, we described the basic theory of oxide charges and interface trap density. Experimental measurement results and conclusion will be described in Sect. 10.3. Comparison of experimental data with Synopsys TCAD device simulation are shown in Sect. 10.4.

© Springer Nature Switzerland AG 2019
A. K. Srivastava, *Si Detectors and Characterization for HEP and Photon Science Experiment*, https://doi.org/10.1007/978-3-030-19531-1_10

10.2   Design of Gated Diode and Method

The Gate controlled diode test field structures of 0.404 mm^2 × 300 µm^2 are used for the extraction of microscopic parameters. The C/V$_g$ measurement is carried out by applying a dc voltage and a small ac signal to the gate contact (2 and 3 gate rings are connected) which drives the device from accumulation, depletion and into inversion [1, 2] and the current-voltage (I/V$_g$) measurement is done by applying constant reverse bias on diode (p$^+$) and varying gate voltage with respect to n$^+$ contact and it is shown in the schematic of the gated diode in Fig. 10.1a. A C/V$_g$ curve is considered high frequency if the probing frequency is 10 kHz. A typical HF C/V$_g$ curve drive from accumulation, depletion and inversion is shown in Fig. 10.1b.

10.2.1   The Ideal MOS Structure

In the real MOS structure (no charges in oxide and no interface trap density), the C/V$_g$ curve of a typical MOS capacitor will exhibit a shift due to the work function

Fig. 10.1 (a) Schematic of the gated diode (five gate rings) structure. Circuit for I/V$_g$ and C/V$_g$ measurements are shown in the left and right part of the structure (here the results are shown for the measurement performed immediately after irradiation). (b) Typical Capacitance-Voltage (C/V$_g$) curve. In panel **b**, C$_{ox}$—oxide capacitance [F], C$_{inv}$—HF inversion capacitance, V$_{fb}$—flat band voltage and C$_{fb}$—flat band capacitance

(a)
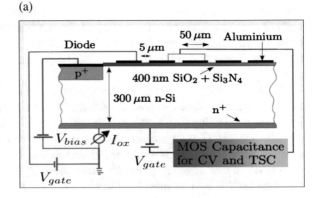

(b)
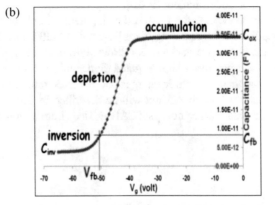

difference (ϕ_{MS}). This is a difference of metal work function (Φ_M) and semiconductor work function (Φ_S).

In Ideal case,

$$V_{fb} = \phi_{MS} = \phi_M - \phi_S \qquad (10.1)$$

In addition to the work function difference, the charges in the insulator and at the Si/SiO$_2$ interface also contributed to the distortion of an ideal C/V$_g$ curve. The presence of these charges is unavoidable in an ideal or real MOS structure. In typical MOS structures the charges are located in various parts of the insulator and the Si/SiO$_2$ interface as shown in Fig. 10.2.

Thus, the C/V$_g$ curve will shift further by V$_{fb}$ (due to N$^{fix}_{ox}$, interface trap density (N$_{it}$ in cm^{-2}) and Φ_{MS}). V$_{fb}$ is given by

$$V_{fb} = \phi_{MS} - \frac{N^{eff}_{ox} qA}{C_{ox}}, \qquad (10.2)$$

where $N_{ox}^{\ eff} = N_{ox}^{\ fix} + N_{it}$ = effective interface charge (cm^{-2}).

V_{fb} = flat band shift due to N$^{eff}_{ox}$ alone.
C_{ox} = Capacitance of the oxide layer (F).
A—Gate area for gated controlled diode.

Fig. 10.2 Charges and their location for thermally oxidized silicon

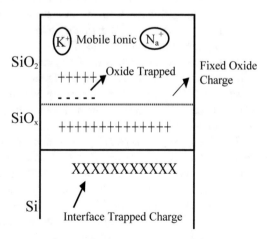

10.2.2 Classification of the Insulator and the Si/SiO₂ Interface Charges

In this section, the basic classification of the insulator and the Si/SiO_2 interface charge that will distort the ideal C/V_g curve will be discussed. There are four general types of charges associated with the Si/SiO_2 system [12]. They are interface trap charge, fixed oxide charge, oxide trap charge, mobile oxide charge.

The interface trap charge N_{it} is the charge, which is due to (1) structural, oxidation-induced defects, (2) metal impurities, or (3) incomplete dangling bonds and absorption of foreign material at the silicon surface. The interface trap charge can be positive or negative charges and located at the Si/SiO_2 interface. Theory has been predicted that each incomplete silicon bond will create an interface trap. The density of the interface trap charge can be reduced when the surface is thermally oxidized.

This is attributed to the bonding of the SiO_2 to the silicon surface atoms. This property of the silicon-dioxide system is unique and is one of the major reasons for using thermally grown SiO_2 for insulating purposes.

Typical values for N_{it} lie in the range of 5×10^9 to 10^{11} cm^{-2}, depending on the orientation of the silicon crystal and on the process history in non-irradiated device and after irradiation, it will increase up to few 10^{12}–10^{13} cm^{-2}. The lowest interface trap density is found on <100> oriented crystal. The interface trap can be positively charged or negatively charged depending whether the trap is acceptor or donor type. An acceptor-type trap becomes negatively charged when it gains an extra electron and becomes neutral when it loses the extra electron. A donor-type trap becomes positively charged when it losses an electron and neutral when it regains the lost electron. It has been found that donor lie in the lower part of band gap and acceptor in the upper part of the band gap of Si.

This increases the slope of the C/V_g curve. Under small signal low frequency, it will respond and it will oscillate about the mean position of the Fermi level and this is capable of increases (High Resistivity Silicon-HRS) and decreases (Low Resistivity Silicon-LRS) interface state density and this will also shift the C/V_g curve with frequency and it depends upon the interface state density.

The fixed oxide charge N^{fix}_{ox} is a positive charge, due primarily to structural defects in the oxide layer less that 25 Å from the Si/SiO_2 interface. This charge is immobile under an electric field. However, it is affected by temperature above 500 °C and by the ambient atmosphere. Typical values for N^{fix}_{ox} in non-irradiated device are in the order of 10^{10}–10^{11} charges per cm^{-2}, depending on process conditions and after irradiation, this will increase up to few 10^{12} cm^{-2}. N^{fix}_{ox} is not affected by the thickness of the oxide. However, it increases when the structure is exposed to high energy radiation.

The density of the fixed oxide charge can be reduced and may be saturate up to some value by high temperature annealing. This fixed oxide charge gives flat band shift and also increases the slope of the C/V_g curve from the ideal MOS C/V_g curve.

The oxide trap charge N_{ot} is due to imperfections throughout the bulk of the oxide layer. It can be positive or negative depending whether holes or electrons are trapped in the bulk of the oxide. Trapping may result from ionising radiation, avalanche injection, or other similar processes and the threshold voltage can be shifted in either direction.

Under ionizing radiation effect, hole traps are positively charges and thus it can be model as fixed positive oxide charges at the SiO_2/Si interface.

The mobile oxide charge N_m is the most significant charge component in the insulator. It is due to ionised impurities such as sodium (Na^+) and, to a less extent, potassium (K^+) and lithium (Li^+). The alkali ions, and in particular sodium, are troublesome and difficult to control. Sodium is a widely distributed impurity in many metal and laboratory and is easily transmitted by human contact. It can migrate in silicon dioxide even at room temperature. Negative ions such as Cl^-, F^- may also be present in the insulator but are not believed to be mobile at temperatures below roughly 500 °C.

The mobile oxide charge gives hysterics effect [2].

The effect of these various charges associated with the SiO_2/Si system can be determined by using different C/V_g measurement. For instance, HF C/V_g measurement is used to determine the effective interface charge N_{ox}^{eff}, by assuming all the insulator charge located at the silicon surface. The effective interface charge will cause the similar shift in the CV curve as that of the actual insulator charge of unknown distribution. Whereas the Quasi-static CV method can be also used to determine the distribution of the interface trapped charge throughout the band gap.

10.2.3   Theoretical Calculation

In this section, we described the most commonly used equations for the understanding and calculation of parameters for MOS capacitors for C/V_g characterization. Under HF signal (f ~ 1 MHz), interface trap will give no respond to the small signal ac excitation and thus, high frequency capacitance can be calculated by following expression;

$$C_{HF,inv} = \frac{C_{ox}C_D}{C_{ox} + C_D}   \quad (10.3)$$

where C_D is depletion layer capacitance.

$$C_{HF,inv} = \frac{1}{\dfrac{1}{C_{ox}} + \dfrac{x_{d,T}}{\varepsilon_S}}   \quad (10.4)$$

where $x_{d,T}$ [1] is the maximum depletion width in MOS capacitor (electric field at $x_{d,T} = 0$). With the comparison of above equation, C_D can be determined.

The interface generation current (I_{ox}) is caused by the surface generation/recombination of free charge carrier with interface trap,

$$I_{ox} = q_0 n_i S_0 A \qquad (10.5)$$

where q_0 is elementary charge, n_i is intrinsic carrier concentration.

Assuming a homogeneous distribution of interface states across the band gap, surface recombination velocity (S_o) is given as.

$$S_0 = \sigma_{eff} v_{th} \pi K_B T D_{itmidgap} \qquad (10.6)$$

where σ_{eff} is the effective capture cross-section of charge carrier, v_{th} is the thermal velocity of charge carrier, K_B is Boltzmann constant, T is temperature and D_{it} is interface trap density at mid gap of Si (cm^{-2} eV^{-1}).

Assuming homogeneous distribution of interface trap in the band gap of Si, D_{it} can be replaced by N_{it} and $\Delta E \approx E_g$ or $E_g/2$ (depends upon the range of energy levels where the interface traps (donor/acceptor or both) are active).

$$N_{it} = D_{it} \Delta E \qquad (10.7)$$

The capacitance (C) and conductance (G/ω) curve for non-irradiated and irradiated gate diode can be corrected for series resistance and other interfacial effects from [9].

The corrected capacitance (C_c) and conductance (G_c) is made by following expression;

$$\left.\begin{aligned} a &= G_m - \left(G^2_m + \omega^2 C^2_m\right)R_S \\[6pt] R_S &= \frac{G_{ma}}{G^2_{ma} + \omega^2 C^2_{ma}} \\[6pt] C_c &= \frac{\left(G^2_m + \omega^2 C^2_m\right)C_m}{a^2 + \omega^2 C^2_m} \\[6pt] G_c &= \frac{\left(G^2_m + \omega^2 C^2_m\right)a}{a^2 + \omega^2 C^2_m} \end{aligned}\right\} \qquad (10.8)$$

where C_m, G_m, ω and R_s is the measured capacitance, conductance, frequency and series resistance.

10.2.4 Two Gaussian Distribution of Interface Trap Model for Surface Damage Effect in Si Sensors

From the TDRC measurements, we got the information about the interface trap versus energy level and its parameters like D_{it}, E_c-E_{it}, σ_{it} (width of Gaussian profile distribution of interface trap). Figure 10.3 describes the two Gaussian distribution of interface trap model for TCAD simulation.

For this, the required distribution of density of states (DOS) for two Gaussian distribution of interface trap in the band gap of Si is given by,

$$DOS = \Big[D_{it,1} e^{-0.5\left(\frac{E-E_{it,1}}{\sigma_{it,1}}\right)^2} + D_{it,2} e^{-0.5\left(\frac{E-E_{it,2}}{\sigma_{it,2}}\right)^2} \Big] \tag{10.9}$$

this will be implemented into Synopsys TCAD device simulator to reproduce the measurements and simulation of radiation damage effects by 12 keV X-rays.

Fig. 10.3 Two Gaussian distribution of interface trap model for TCAD simulation of surface damage effect in Si sensors

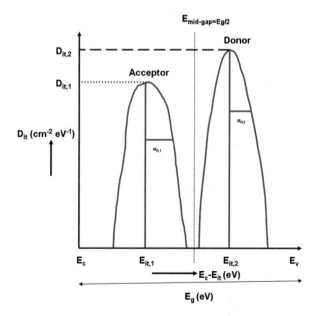

10.3   Results and Conclusion

A lot of progress has already been made in order to understand the radiation hardness of silicon sensors for XFEL environment and the important observation has been found that after a few MGy of irradiations, saturation of charge occurs [4].

It is interesting to see the behaviours of capacitance and conductance versus frequency up to 5 MGy irradiation doses [4].

Here, we present our new result and views about the capacitance and conductance before and after irradiation.

10.3.1   Non-irradiated Result: C/V_g as Function of Frequency

In non-irradiated gated diode, we have found that the decrease of oxide related capacitance (C_{ox}) with increase of frequency in Fig. 10.4a. This is due to uncorrected series resistance of the bulk (R_{SB} for high frequency model [21]). We have performed the correction in capacitance using Eq. (10.7) only at 10 kHz and 800 kHz, it has been found that after correction, there is no change in C_{ox} and no shift in the C/V_g (co change in flat band voltage) curve with increasing frequency. This is due to low interface trap concentration.

10.3.2   Non-irradiated Result: G/V_g as Function of Frequency

The G/V_g as a function of frequency is shown in Fig. 10.4b. It is observed that G increases with frequency. This shows that there is no peak in the G/V_g curve at high frequency so we have to correct the series resistance in the accumulation region of MOS using Eqs. (10.8) and (10.9).

Figure 10.5a, b show the corrected conductance at 10 and 800 kHz. There is peak observed at certain bias and it is increases with frequency. The peak of the G/V_g curve depends upon the distribution of the charge carrier at the Si-SiO$_2$ interface and it is seen that peak is at around flat band voltage. This also confirms from simulation [Sri].

10.3.3   Irradiated Result: C/V_g, G/V_g as Function of Frequency

In our previous results [3], we have shown the shift of the C/V_g curve with frequency for an irradiated gate diode [4]. It is interesting to see in Fig. 10.6, there is a peak of the C/V_g curve at around flat band voltage at 10 kHz with irradiation doses from 0.5 MGy to 10 MGy, this is due to increase of the charge distribution at Si-SiO$_2$

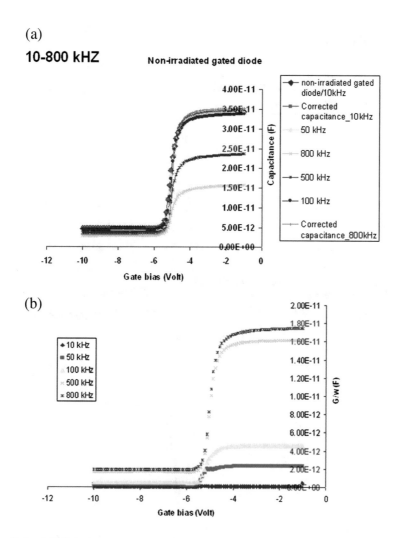

Fig. 10.4 (**a**) C/V$_g$ as function of frequency for non-irradiated gated diode. (**b**) G/V$_g$ as function of frequency for non-irradiated gated diode

interface. But for 10 MGy, there is decrease of the peak and flat band voltage and this is because of the decreases of the charge distribution after 5 MGy [4].

In Fig. 10.7a, we have shown the G/ω/V$_g$ curve for 5 MGy for different frequency. The similar behavior is achieved as per Fig. 10.2, G/ω increases with frequency.

There is shift in the peak of the conductance curve observed towards lower gate voltage with increasing frequency. The conductance peak is visible at lower frequency but at higher frequency like at 800 kHz, it is not clear. Therefore, we corrected the conductance (shown in Fig. 10.7b) for 10 and 800 kHz, it can be

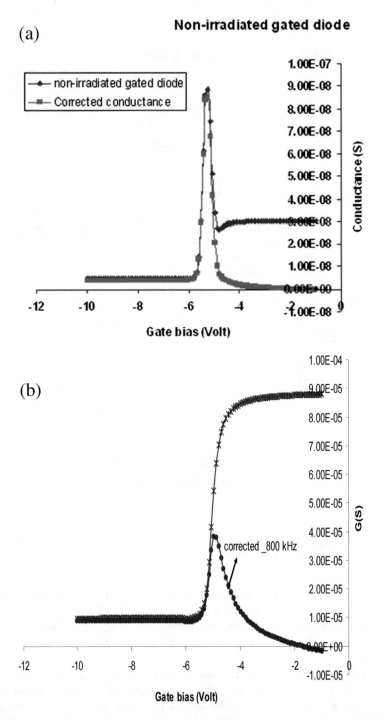

Fig. 10.5 (**a**) Corrected conductance as function of gate bias at for non-irradiated gated diode at 10 kHz. (**b**) Corrected conductance as function of gate bias at for non-irradiated gated diode at 800 kHz

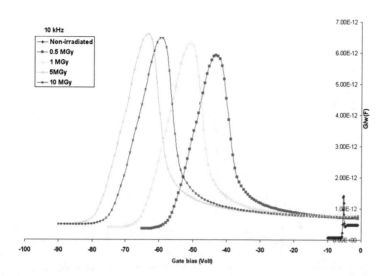

Fig. 10.6 Conductance (G/ω) as function of gate bias at for different irradiated gated diode at 10 kHz

seen the peak is clear for both frequency and shift of the peak towards lower gate voltage.

In Fig. 10.7b, we have only corrected the capacitance of 5 MGy irradiated gate diode at 10 kHz and 800 kHz and two important things is observed from the real corrected capacitance.

There is no change in uncorrected and corrected capacitance is observed for low frequency of 10 kHz. Thus, this can be optimum frequency for the gated diode measurement whereas at high frequency of 800 kHz, there is change in uncorrected and corrected capacitance at 800 kHz. It can be seen that in Fig. 10.8, a peak in the accumulation region observed, may be this is due to the mechanism that charge transfer through the Si-SiO₂ interface or we can correct the capacitance from model [21].

10.4   Comparison of Simulation and Experiments

There is good agreement observed between simulation and experiment for non-irradiated MOS test structure and already shown detailed results in [Sri] and also reported first set of results on comparison with 0.5 MGy irradiated MOS test structure as function of frequency [21]. Now, we have new set of experimentally measured microscopic parameters which is based on two Gaussian distribution of interface trap model for surface damage effect in Si sensors for the comparison of test structures results and simulation and prediction of the Si sensors in the XFEL environment and this is underway.

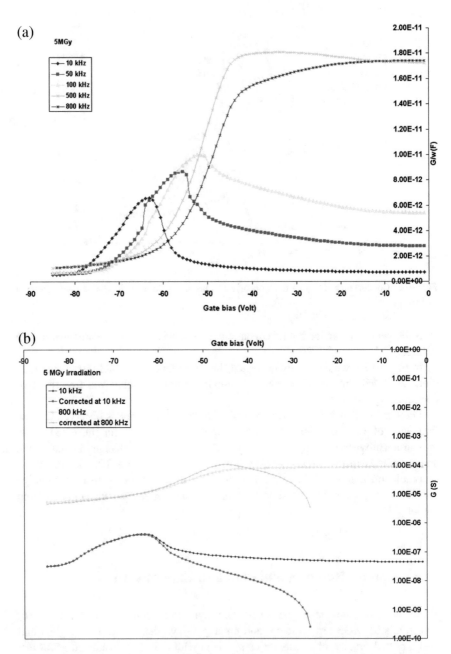

Fig. 10.7 (**a**) Conductance (G/ω) at different frequency as a function of gate bias for 5 MGy irradiated gated diode. (**b**) Conductance (G) at 10 and 800 kHz frequency as a function of gate bias for 5 MGy irradiated gated diode

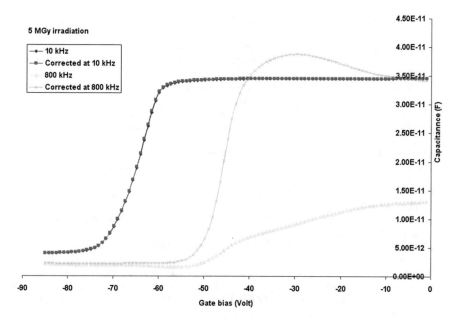

Fig. 10.8 Corrected capacitance at 10 and 800 kHz as function of gate bias for 5 MGy irradiated gated diode

This experience is used in the design of radiation tolerant segmented detectors p^+n silicon pixel sensor for the AGIPD at EuXFEL.

Acknowledgement The author specially would like to thank the H. Perrey, University of Hamburg for providing data for analysis and E. Fretwurst, R. Klanner for useful discussion and suggestions and constant academic supports during my comparison of simulation and data. The authors would also like to thank the XFEL company for support and also would like to thank to the peoples involved in the development of AGPID for XFEL experiment from DESY (Deutsches Elektronen Synchrotron), PSI (Paul Scherer institute), Switzerland and University of Bonn, Germany for constant interest and support. This work was profited from the infrastructure grant of the Helmholtz Alliance "Physics at the Terascale".

References

1. The European X-Ray Laser Project *XFEL*, http://xfel.desy.de/
2. Becker, J., et al., on behalf of the AGIPD Consortium: Plasma effects in silicon detectors for the European XFEL and their impact on sensor performance. Nucl. Instr. Methods Phys. Res. A **615**(2), 230–236 (2010)
3. Klanner, R., et al.: Study of high-dose X-ray radiation damage of silicon sensors. Nucl. Instr. Methods Phys. Res. A. **732**, 117–121 (2013)
4. Fretwurst, E., et al.: Study of the radiation hardness of silicon sensors for the XFEL, poster presented in IEEE NSS 2008, Dresden, Germany, N30–400

5. Srivastava, A. K., et al.: Simulation of MOS capacitor for C/V$_g$ characterization (internal note 2010)
6. Tataroğlu, A., et al.: Analysis of electrical characteristics of Au/SiO2/n-Si (MOS) capacitors using the high-low frequency capacitance and conductance method. Microelectron. Eng. **85**, 2256–2260 (2008)
7. Koukab, A., et al.: An improved high frequency method C-V method for interface state analysis on MIS structures. Solid State Electron. **41**(4), 635–641 (1997)
8. Beck, D., et al.: IEEE Trans. Electron. Dev. **44**(7pp), 1091–1101 (1997)
9. Schroder, D.K.: Semiconductor material and device characterization. Wiley, Hoboken, NJ (2006)
10. Schroder, D.K.: Advanced MOS devices. Addison-Wesley, Reading, MA (1987)
11. Kuhn, M.: A quasi-static technique for MOS CV and surface state measurements. Solid State Electron. **13**, 873–885 (1970)
12. Ooi Chee Pun MSc. Report, MOS capacitor-measurement and parameter extraction. Queen's University of Belfast (1997)
13. Zaininger, H., Heiman, F.P.: The CV technique as an analytical tool. Part 1. Solid-State Technol. **May**, 49–56 (1970)
14. Zaininger, H., Heiman, F.P.: The CV Technique as an Analytical Tool. Part 2. Solid-State Technol. **June**, 49–56 (1970)
15. Zeigler, K., Klausmann, E., Kar, S.: Determination of the semiconductor doping profile right ups to its surface using the MIS capacitor. Solid State Electron. **18**, 189–198 (1975)
16. Mego, T.J.: Guidelines for interfacing CV data. Keithley Instruments, Cleveland, OH. Solid-State Technol. (May, 1990)
17. Berglund, C.N., member IEEE: Surface states at stream-grown silicon-silicon dioxide interface. IEEE Trans. Electron Dev. **ED-13**(10) (Oct 1966)
18. Pierret, R.F.: Field effect devices. Addison-Wesley, Reading, MA (1990)
19. Grove, A.S.: Physics and technology of semiconductor devices. Wiley, New York (1967)
20. Nicollian, E.H., Brews, J.R.: MOS physics and technology. Wiley, New York (2000)
21. Srivastava, A.K., et al.: Numerical modelling of the frequency behaviours of the irradiated MOS test structure (DESY Internal note)

Chapter 11
Si Detector for HEP and Photon Science Experiments: How to Design Detectors by TCAD Simulation

11.1 Si Detectors for HEP Experiments

Si detector is widely used in HEP experiments for e.g. in CMS experiment at LHC at CERN.

A particle emerging from the high-energy particle collision at LHC and travelling outwards will first encounter the tracking system of the CMS experiment, made of **Si pixels and Si strip detectors**. These accurately measure the positions of passing charged particles allowing scientist to reconstruct their tracks. Charged particles follow spiraling paths in the CMS magnetic field and the curvature of their paths reveal their particle momenta.

The energies of the charged particles will be measured in the next layer of the CMS detector, the so-called electromagnetic calorimeters. Electrons, photons and jets (sprays of particles produced by quarks) will all be stopped by the calorimeters, allowing their energy to be measured.

The first calorimeter layer is designed to measure the energies of electrons and photons with great precision. Since these particles interact electromagnetically, it is called an electromagnetic calorimeter (ECAL).

Particles that interact by the strong force, hadrons, deposit most of their energy in the next layer, the hadronic calorimeter (HCAL). The only known particles to penetrate beyond the HCAL are muons and weakly interacting particles such as neutrinos. Muons are charged particles, which are then tracked further in dedicated muon chamber detectors. Their momenta are also measured from the bending of paths in the CMS magnetic field. Neutrinos, however, are neutral and since they hardly interact at all they will escape detection. Their presence can nevertheless be inferred. By adding up the momenta of all the detected particles, and assigning the missing momentum to the neutrinos, CMS physicists will be able to tell where these particles were.

Particles travelling through CMS thus leave behind characteristic patterns, or 'signatures', in the different layers, allowing them to be identified. The data is passed

© Springer Nature Switzerland AG 2019
A. K. Srivastava, *Si Detectors and Characterization for HEP and Photon Science Experiment*, https://doi.org/10.1007/978-3-030-19531-1_11

from the CERN to the centres around the World where analysts then reconstruct the "event" and the presence of any new particles can be gathered.

The Design & Technology of Si Strip (p in n and n in p type with p-stop and p-spray on different Si materials-SFz, DOFz, MCZ, Cz) & Pixel Detectors (p in n and n in n, or 3D & CMOS Monolithic Active Pixel Sensor: CMOS MAPS) has started to achieve our Milestones for the Phase2 Upgrade at HL-LHC (started in 2026) as discussed in our previous chapters.

11.2   Si Detectors for Photon Science Experiments

Photon science encompasses all aspects of creating, measuring and using light (ultra short wavelength X-rays) for **science**. **Photon science** allows us to discover new things about the properties of all kinds of physical and biological matter (structures of proteins in order to better understand the causes of illness and develop better treatments), enabling us to answer questions of what things are and how and why they work. May scientists are using X rays to see how the atoms are arranged in nanomaterial's samples.

Conventional X-ray tubes like those used in hospitals are generally not suitable for today's scientific applications, because the X-rays they generate are too weak for the experiments. Particle accelerators, on the other hand, generate extremely powerful and focused radiation. These X-ray beams are so intense that they can reveal even the finest details—for example, the tiniest cracks and pores in a turbine blade, minute impurities in a semiconductor, or the positions of individual atoms in a protein molecule. Moreover, when researchers fire extremely short X-ray flashes at various samples, they are able to observe ultrafast processes such as those that occur in a chemical reaction.

The DESY campus in Hamburg hosts some of the world's best light sources: the PETRA III storage ring generates brilliant X-ray light for various experiments; the FLASH free-electron laser produces ultrashort laser pulses in the soft X-ray range; and the European XFEL X-ray laser is a true super microscope. These facilities make DESY the world's leading centre for research with X-rays.

Nowadays, silicon detectors are used in practically all particle physics experiments at accelerators; they also play a central role in photon science at synchrotron-radiation sources and free-electron lasers, and find many applications in other fields of science, as well as in medicine and industry.

Figure 11.1 shows the MYTHEN detector (Si strip Detector module) at the powder diffraction station at SLS, PSI, Switzerland.

Figure 11.2 shows the module used in the MYTHEN detector at the SLS, PSI, Switzerland.

The following design and parameters are used as follows: Si-strip sensors thickness (320 micron), 1280 strips (8 mm long strips with 50 micron pitch), counting rate ($>2 \times 10^5$ per strips), max. number of counts 24 bits (16, 777, 216), energy range (5 keV (90%)–30 keV (8%)), and frame rate (25 Hz (24 bit)–500 Hz (4 bit)).

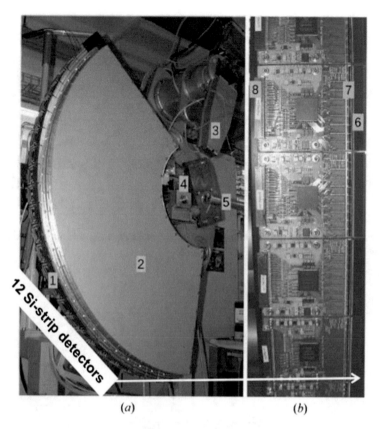

Fig. 11.1 (**a**) Photograph of the MYTHEN detector installed at the powder diffraction at the SLS, PSI, Switzerland and (**b**) a zoom on the device modules building the detector. The numbers indicate the main elements of interest: (1) MYTHEN detector layer; (2) He-filed box behnd which is fixed the DAQ (Data Acquisition System); (3) analyzer crystal detector; (4) center of the diffractometer; (5) beampipe; (6) silicon microstrip detector; (7) front-end electronics; (8) connector to the DAQ [1]

There are varieties of detectors with different designs and working principle mainly on the side-ward depletion that are also used in photon science experiment: CCD (Charge Coupled Devices), Hybrid Pixel Detector (charge integrating, X ray counting, X ray imaging) shown in Fig. 11.3, DEPFET etc.

Now, an X-ray imager AGIPD (Adaptive Gain Integrating Pixel Detector) detector at DESY for the EuXFEL (European X-ray Free-Electron Laser was installed and commissioned in August 2017 (Fig. 11.4). It is a fast, low-noise integrating pixel detector, with an adaptive gain switching amplifier per pixel. The AGIPD at EuXFEL is using hybrid p^+n Si pixel sensor (500 micron thick, 200 micron $\times$ 200 micron) as discussed in our earlier Chap. 4 in detail. The sensor is design to work up to 900 V up to 1 GGY X-ray dose and that AGIPD system can store images up to 352 in the burst mode while runnng at up to frame rate of 6.5 MHz. The dynamic range is $1 \rightarrow 10^4$ photons pixel^{-1} pulse^{-1} (12 keV) by adaptive gain switching.

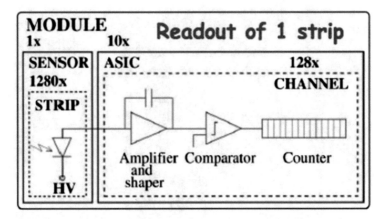

Fig. 11.2 Module (Si strip Sensor and ASIC) used in the MYTHEN detector at the SLS, PSI, Switzerland [1]

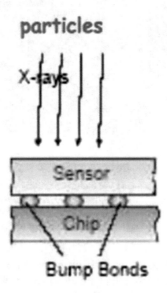

Fig. 11.3 Hybrid pixel Sensors that are bump-bonded with electronics chip

Recent progress in active-edge technology of silicon sensors enables the development of large-area tiled silicon pixel detectors with small dead space between modules by utilizing edgeless sensors. Such technology has been proven in successful productions of ATLAS and Medipix-based silicon pixel sensors by a few foundries. However, the drawbacks of edgeless sensors are poor radiation hardness for ionizing radiation and non-uniform charge collection by edge pixels.

There are a detector roadmap given at DESY in the Fig. 11.5.

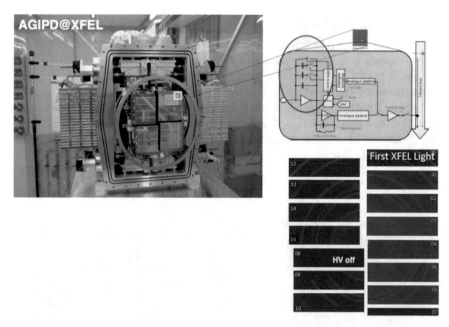

Fig. 11.4   AGIPD at DESY EuXFEL

Detector Roadmap

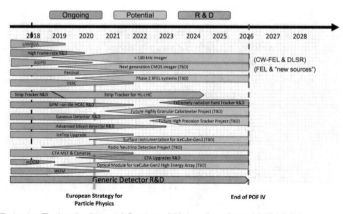

Detector Technologies and Systems| Heinz Graafsma | MT | DTS

Fig. 11.5   Detector roadmap at DESY, Hamburg, Germany (pic credit to H. Graafsma)

Here, we have shown the way to designing of the detectors for the experiments as per the specification provided by the experiments.

11.3   Designing Steps

In this section, here we have taken strip & p^+_n pixel detector design for HEP and Photon Science Experiments and here, we will learn how to design the detectors using Synopsys TCAD simulation.

For any detector design simulation, there are two main parts that can be used to simulate using Synopsys TCAD Device simulation, (I) Inner structure (Fig. 11.6a), (II) Outer part of detector design structure near to cur edge or scribe line of the detector (Fig. 11.6b).

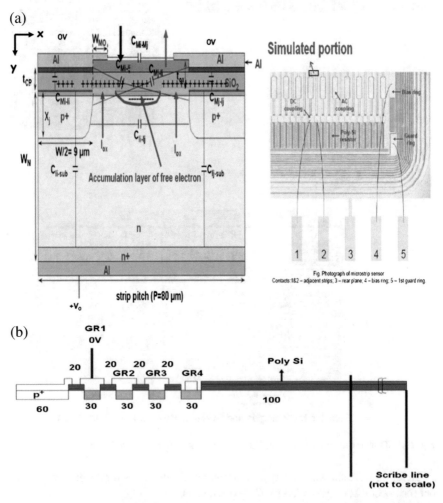

Fig. 11.6 (a) Cross-section of p in n AC coupled Si strip detector (inner structure) as an example. (b) Layout of near to cut edge Si detector design with a MGR (multiple guard rings, orange filled p+), p+ is last strip of each side of detector as an example

The TCAD simulations were performed for a temperature of 293 K (20 °C) using the drift-diffusion physical model. The minority carrier lifetime (generation) in the bulk was assumed to be 1 ms, the charge carrier mobility was modeled doping dependent with a degradation at the Si-SiO$_2$ interface, and for the avalanche process the van Overstraeten—de Man model with the default parameters [2] was used in the present simulation. To take into account possible tunneling effects in very high E-field regions the band-to-band tunneling model of Hurkx [3] was also switched on.

On top of the SiO$_2$ (brown colour in Fig. 11.6a) Neumann boundary (NB) conditions (zero normal component of the E-field) were used. The results from reference [4] imply, that the boundary conditions on top of the oxide separating p$^+$ implants at the same potential change with time from Neumann to Dirichlet (constant potential). The corresponding time constants vary between minutes and days, depending on environmental conditions like relative humidity. It was decided to use Neumann boundary conditions (NBC), as simulations have shown, that they result in a lower breakdown voltage after X-ray irradiation. They are also considered to be valid for the sensor operation in vacuum or in a very dry ambient [5].

For any p+n or n$^+$p Si Si strip and p+n Pixel sensor for any design of experiments (DoE), this is really important to know the first the specification of the Si detectors and requirement of the experiments. Usually, the detector is an abrupt pn junction diode, which is usually operated in the reverse bias mode in order to get the following things:

(a) Low leakage current i.e. generation leakage current
(b) High detection volume
(c) Less junction capacitance, this can be less in the total detector capacitance at the input of the preamplifier for the thick detectors (200–500 μm).

The electrical requirement of the detector is as follows for all types of (Strip/ Pixel) detectors:

 (i) High avalanche breakdown voltage as per the radiation level/photon science (dynamic range) of the detector
(ii) Low leakage current
(iii) Low interstrip/inter-pixel capacitance
(iv) High charge collection with less charge trapping via deep traps or e-h plasma due to zero E-field region in the X ray detector just below the electron accumulation layer

In previous chapters, we have discussed the physics & technology of Si detectors and techniques to improve V_{BD} of the detectors in detail. The effects of bulk and surface damage in Si detectors have been clearly presented.

11.3.1   Designing Using Synopsys TCAD Simulation

For any design of p$^+$n AC coupled Si Strip Detector for e.g. of .643 cm$^2 \times$ 300 μm, 98 strips, the following process and device parameters are suggested to design: Junction depth (X$_j$) = 1 μm, coupling oxide (t$_{cp}$) = 150 μm, field oxide (t$_{ox}$) = 350 μm, width of metal overhang (W$_{MO}$) = 2.5 μm, width of n+ = 1 μm, and width of Al = 1.2 μm. We must first calculate the following things (see Chap. 2 for reference):

(i) Resistivity of Si wafer at 300 K for known doping concentration (N$_D$ for n-type bulk Si = 8.1 $\times$ 10^{11} cm^{-3})
(ii) Full depletion voltage (V$_{FD}$) at given thickness of n-type Si wafer (300 μm)
(iii) Junction capacitance of single p+ strip (Area of one p+ strip = total detector area/number of strips, for 2D simulation, length of strip = 1 μm) at given thickness of n-type Si wafer
(iv) Generation lifetime from experimental current-voltage measurement in non-irradiated Si strip detector at 20 °C of V$_{FD}$

Given as per the specifications provided to design detectors, width (w) and strip pitch (p) (see in Fig. 11.6a) should be noted down for the designing in TCAD. Let us take from Fig. 11.6, w = 18 μm, and p = 80 μm. For the TCAD simulation, we mainly use the Linux OS for the execution.

Firstly, we should start Synopsys TCAD (sdevice simulation) in your computer screen. Let us open a terminal on the screen and type sde & see menu bar on top of the screen (Fig. 11.7 Synopsys TCAD framework on your screen; → **File Edit View Draw Mesh Device Contacts Help**). There are a number of steps that are given below to design the detectors using sdevice TCAD.

(i) Create rectangle sheet region, take Silicon material as a first layer of the detector Go on Draw, Auto region Naming (Tick), → Exact coordinates→ XY, Add rectangle ---- x1 = 0   y1 = 0   x2 = 80   y2 = 300
Enter region name-Substrate
Zoom in
(ii) Add rectangle ----x1 = 9   y1 = 0   x2 = 71   y2 = −0.3
Enter region name-Oxide
(iii) Box preparation of p
x1 = 0   y1 = 0   x2 = 9   y2 = 1
Enter region name-p1
x1 = 71   y1 = 0   x2 = 80   y2 = 1
Enter region name-p2
(iv) x1 = 0   y1 = 299   x2 = 80   y2 = 300
Enter region name-n+
(v) Select Vertex
Edit-2 D edit tool
Set filet-0.8
(vi) Contacts Set

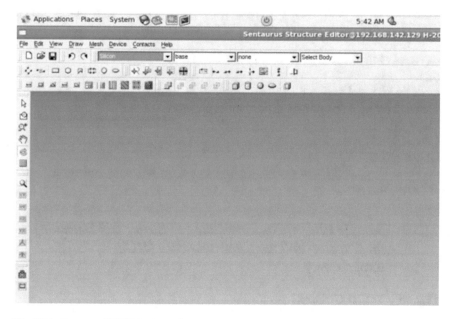

Fig. 11.7 Synopsys TCAD framework on your screen

Contact Name & Set
Pcontact1
Pcontact2
Ncontact
Select edge, Contacts Set for all
(vii) Device
I. Constant Profile Placement
Region-Substrate
$N_D = 1 \times 10^{12} \text{ cm}^{-3}$
Add placement
II. Constant Profile Placement
Region- n+
$N_D = 1 \times 10^{19} \text{ cm}^{-3}$
Add placement
III. Device-Analytical Profile Placement
Peak doping concentration(P) $= 1 \times 19 \text{ cm}^{-3}$, junction depth at $-1 \times 14 \text{ cm}^{-3}$
(viii) Mesh-Define RelEval Window
Or Mesh Refinement Placement
Region wise
Max: $X = 0.02$ $Y = 0.02$
Min: $X = 0.02$ $Y = 0.01$

Then click on Mesh to create a "Build Mesh". Here it is important to fine the grid in the junction and oxide than the rest of the layer of the detectors (see Appendix A.1.1).

(**ix**) Save Grid file as a

SiD

(seen by SiD_msh.tdr) in the current directory after run by sdevice pn_des.cmd

That will set up a few of files in the current directory.

Plot="SiD"

Current="SiD"

Output="SiD"

(**x**) For analysis of detector design, we have to use the following command;

svisual& SiD_des.plt

11.3.2 Sample File for Non-irradiated Si Strip/Pixel Detector for Current-Voltage Characteristics (HEP Experiment)

11.3.2.1 TCAD CODING

- **Sample pn_des.cmd file for non-irradiated Si strip/pixel detector (p in n type) for current-voltage (I-V) characteristics at 20 °C (HEP experiment)**

```
Electrode{
  { name="Pcontact1"        voltage= 0.0 Areafactor=7.3e4 }
  { name="Pcontact2"        voltage=0.0}
  { name=" Ncontact "       voltage=0.0}
}
```

∗Area factor can be used to normalized area of strip or pixel∗
∗ File Description--∗

```
File{
  Grid = "SiD_msh.tdr"
  Parameter="models.par"
  Parameter="Oxide.par"
  Plot="SiD"
  Current="SiD"
  Output="SiD"
}
```

```
∗----------------------------------------------------------------∗
Plot {
eDensity hDensity eCurrent hCurrent
Potential SpaceCharge ElectricField

eMobility hMobility eVelocity hVelocity
eQuasiFermi hQuasiFermi Potential
SRHRecombination Auger AvalancheGeneration
```

```
Doping DonorConcentration AcceptorConcentration
Current
surfaceRecombination
eIonIntegral hIonIntegral
eAlphaAvalanche hAlphaAvalanche
eTrappedCharge hTrappedCharge
HeavyIonChargeDensity
}


Physics{
Temperature=293
Mobility( DopingDep HighFieldSaturation Enormal
CarrierCarrierScattering (ConwellWeisskopf))
EffectiveIntrinsicDensity( BandGapNarrowing (OldSlotboom) )
Recombination ( SRH ( DopingDependence ) SRH(TempDependence)
eAvalanche(CarrierTempDrive) havalanche (Eparallel))
ComputeIonizationIntegrals(WriteAll)
Add2TotalDoping Tunneling (Hurkx))
}


Physics(
 MaterialInterface ="Silicon/Oxide") {
  charge (Conc=1e10)
Recombination(surfaceSRH)
}
```

* For non-irradiated Si detector, this charge will around 1×10^{10} to 1×10^{11} cm^{-2} and that will depend on <100>, or <111> orientation. This fixed oxide charge (N_{ox}) value can be extracted from capacitance-voltage measurement on non-irradiated n-MOS test structure that is having similar thickness of dielectric passivation layer or the vendor can give with Si wafers. Or plot $1/C^2$ versus Bias voltage for (interface trap density in cm^{-2} ($N_{it}=0$), and $N_{ox}=0$ and values from vendors). For the calculated value of V_{FD}, we can estimate the N_{ox} in the design. *

```
*------------------------------------------------------------------*
Math{
  Extrapolate
  Derivatives
  NewDiscretization
  NotDamped=50
  Iterations=10
  BreakATIonIntegral
}


Solve {
Poisson
Coupled{ Poisson Electron Hole }
Quasistationary (
    InitialStep=0.1 Increment=1.3
    MaxStep =0.1 MinStep = 1e-15
    Goal{ Name="Ncontact" voltage =200}
  ){ Coupled {Poisson Electron Hole} }
}
```

* The voltage on Ncontact should be higher than full depletion voltage+50 V→ 150 V for over depleted Si detector.

The following electrical parameters can be extracted from aforesaid TCAD program but not limited:

(a) Electrostatic potential
(b) E-Field as a function of the detector depth and along x-axis of the detector
(c) Current-voltage characteristics (current on Pcontact1/Pcontact2 as a function of Ncontact voltage)

11.3.3   Sample File for Non-irradiated Si Strip/Pixel Detector for Capacitance-Voltage (C-V) Characteristics (HEP Experiment)

11.3.3.1   TCAD CODING

- **Sample pn_des.cmd file for non-irradiated Si strip/pixel detector (p in n type) for capacitance-voltage characteristics at 20 °C (HEP experiment)**

```
Device GD {

Electrode{
  { name="Pcontact1"        voltage= 0.0 Areafactor=7.3e4 }
  { name="Pcontact2"        voltage=0.0}
  { name=" Ncontact "       voltage=0.0}
}
```

Area factor can be used to normalized area of strip or pixel
```
* File Description--------------------------------------------------*

File{
  Grid = "SiD_msh.tdr"
  Parameter="models.par"
  Parameter="Oxide.par"
  Plot="SiD"
}


*------------------------------------------------------------------*
Plot {
eDensity hDensity eCurrent hCurrent
Potential SpaceCharge ElectricField

eMobility hMobility eVelocity hVelocity
eQuasiFermi hQuasiFermi Potential
SRHRecombination Auger AvalancheGeneration
Doping DonorConcentration AcceptorConcentration
```

```
Current
surfaceRecombination
eIonIntegral hIonIntegral
eAlphaAvalanche hAlphaAvalanche
eTrappedCharge hTrappedCharge
HeavyIonChargeDensity
}


Physics{
Temperature=293
Mobility( DopingDep HighFieldSaturation Enormal
CarrierCarrierScattering (ConwellWeisskopf))
EffectiveIntrinsicDensity( BandGapNarrowing (OldSlotboom) )
Recombination ( SRH ( DopingDependence ) SRH(TempDependence)
eAvalanche(CarrierTempDrive) havalanche (Eparallel))
ComputeIonizationIntegrals(WriteAll)
Add2TotalDoping Tunneling (Hurkx))
}


Physics(
 MaterialInterface ="Silicon/Oxide") {
  charge (Conc=1e10)
Recombination(surfaceSRH)
}
```

* For non-irradiated Si detector, this charge will around $1x10^{10}$ to $1x10^{11}$ cm^{-2} and that will depend on <100>, or <111> orientation. This fixed oxide charge (N_{ox}) value can be extracted from capacitance-voltage measurement on non-irradiated n-MOS test structure that is having similar thickness of dielectric passivation layer or the vendor can give with Si wafers. Or plot $1/C^2$ versus Bias voltage for (interface trap density in cm^{-2} ($N_{it}=0$), and $N_{ox}=0$ and values from vendors). For the calculated value of V_{FD}, we can estimate the N_{ox} in the design. *

```
*----------------------------------------------------------------*
Math{
  Extrapolate
  Derivatives
  NewDiscretization
  NotDamped=50
  Iterations=10
  BreakATIonIntegral
}


File {
Output = "SiD"
ACExtract="SiD"
}


{
GD trans(Pcontact1 Pcontact2 Ncontact)
Vsource_pset Vp1   (Pcontact1 0) {dc=0 }
Vsource_pset Vn    (Ncontact 0) {dc=0 sine=(0 0.1 0.01meg 0 0)}
Vsource_pset Vp2   (Pcontact2 0) {dc=0 }
}
```

* Frequency at fixed 10 KHz and may be it will vary from start to end.

```
Solve {
 Poisson
 Coupled{ Poisson Electron Hole }
# Ramp Vn and apply ac on Ncontact
Quasistationary (
     InitialStep=0.1 Increment=1.3
     MaxStep =0.1 MinStep = 9e-5
     Goal{ Paramater=Vn.dc Voltage= 0}
     ){ Coupled {Poisson Electron Hole}}
Quasistationary (
     InitialStep=0.01 Increment=1.3
     MaxStep =0.02 MinStep = 9e-5
     Goal{ Paramater=Vn.dc Voltage= 200}
)
{
ACCoupled (
StartFrequency =1e4 EndFrequency=1e4
NumberOfPoints=1 Decade
Node(Pcontact1 Ncontact Pcontact2 ) Exclude (Vp1 Vn Vp2)
)
{Poisson Electron Hole}}

 }
```

* **The voltage on Ncontact should be higher than full depletion voltage+50 V→ 150 V for over depleted Si detector.**

The following electrical parameters can be extracted from aforesaid TCAD program but not limited:

(a) Junction capacitance between Pcontact1 and Ncontact electrode at V_{FD} (pF/μm)
(b) Inter strip & electrode capacitances between different electrodes (see reference [6])
(c) Capacitance-voltage characteristics to get Junction capacitance at V_{FD}
(d) $1/C^2$ versus bias voltage to get V_{FD}

11.3.4   Sample File for Surface Irradiated p+n Si Pixel Detector (Inner/Outer) for Current-Voltage Characteristics

11.3.4.1   TCAD CODING

- **Sample pn_des.cmd file for surface-irradiated Si strip/pixel detector (p in n type) for current-voltage (I-V) characteristics at 20 °C (HEP experiment)**

```
Electrode{
  { name="Pcontact1"           voltage= 0.0 Areafactor=7.3e4 }
  { name="Pcontact2"            voltage=0.0}
  { name=" Ncontact "          voltage=0.0}
}
```

Area factor can be used to normalized area of strip or pixel
* File Description--*

```
File{
  Grid = "SiD_msh.tdr"
  Parameter="models.par"
  Parameter="Oxide.par"
  Plot="SiD"
  Current="SiD"
  Output="SiD"
}
```

--
```
Plot {
eDensity hDensity eCurrent hCurrent
Potential SpaceCharge ElectricField

eMobility hMobility eVelocity hVelocity
eQuasiFermi hQuasiFermi Potential
SRHRecombination Auger AvalancheGeneration
Doping DonorConcentration AcceptorConcentration
Current
surfaceRecombination
eIonIntegral hIonIntegral
eAlphaAvalanche hAlphaAvalanche
eTrappedCharge hTrappedCharge
HeavyIonChargeDensity
}
```

```
Physics{
Temperature=293
Mobility( DopingDep HighFieldSaturation Enormal
CarrierCarrierScattering (ConwellWeisskopf))
EffectiveIntrinsicDensity( BandGapNarrowing (OldSlotboom) )
Recombination ( SRH ( DopingDependence ) SRH (TempDependence)
eAvalanche(CarrierTempDrive) havalanche (Eparallel))
ComputeIonizationIntegrals(WriteAll)
Add2TotalDoping Tunneling (Hurkx))
}
```

```
Physics(
 MaterialInterface ="Silicon/Oxide") {
  charge (Conc=1e12)
  Traps (Donor Level EnergyMid=0 fromMidBandGap conc=1.3e10
eXsection=2.9e-16 hXsection=2.9e-16 ElectricField Tunneling
(Hurkx))
```

```
* Traps (Acceptor level fromCondBand conc=5.5e14 EnergyMid=0.40
eXsection=7e-17 hXsection=7e-17 )*
Recombination(surfaceSRH)
}
```
* Surface interface traps parameters (type of trap, position of trap in
Silicon, capture cross-section of electron and holes) can be extracted
from TDRC (thermally depolarization relaxation measurements on
irradiated MOS capacitor with different X-ray doses or use different
surface recombination velocity for different X-ray doses*
* For irradiated Si detector, this charge will around $1x10^{12}$ to $3x10^{13}$ cm^{-2}
and that will depend on irradiation with X-ray doses . This effective
fixed oxide charge ($N_{ox}+N_{it}$) value can be extracted from capacitance-
voltage measurement on irradiated n-MOS test structure that is having
similar thickness of dielectric passivation layer. Or plot $1/C^2$ versus
Bias voltage for (interface trap density in cm^{-2} (N_{it}=0), and N_{ox}=0 and
values from vendors). For the calculated value of V_{FD}, we can estimate the
($N_{ox}+N_{it}$) in the irradiated detector. *
```
*-----------------------------------------------------------------*
Math{
  Extrapolate
  Derivatives
  NewDiscretization
  NotDamped=50
  Iterations=10
  BreakATIonIntegral
}


Solve {
Poisson
Coupled{ Poisson Electron Hole }
Quasistationary (
    InitialStep=0.1 Increment=1.3
    MaxStep =0.1 MinStep = 1e-15
    Goal{ Name="Ncontact" voltage =200}
  ){ Coupled {Poisson Electron Hole} }
}
```

The following electrical parameters can be extracted from aforesaid TCAD program but not limited:

(a) Electrostatic potential
(b) E-Field as a function of the detector depth, along x-axis of the detector, and E-field just below the electron accumulation layer (EAL)
(c) Current-voltage characteristics (current on Pcontact1/Pcontact2 as a function of Ncontact voltage), here we have to look the difference between non-irradiated and surface irradiated detector for surface current at V_{FD}
(d) e-concentration in the EAL just below the Si-SiO$_2$ interface

* **The voltage on Ncontact should be higher than full depletion voltage+50 V→ 150 V for over depleted Si detector**

11.3.5   Sample File for Surface Irradiated p+n Si Detector (Inner/Outer) for Capacitance-Voltage Characteristics

11.3.5.1   TCAD CODING

- **Sample pn_des.cmd file for surface-irradiated Si strip/pixel detector (p in n type) for capacitance-voltage characteristics at 20 °C (HEP experiment)**

```
Device GD {

Electrode{
  { name="Pcontact1"        voltage= 0.0 Areafactor=7.3e4 }
  { name="Pcontact2"         voltage=0.0}
  { name=" Ncontact "        voltage=0.0}
}
```

∗Area factor can be used to normalized area of strip or pixel∗

```
* File
Description---------------------------------------------------------*

File{
  Grid = "SiD_msh.tdr"
  Parameter="models.par"
  Parameter="Oxide.par"
  Plot="SiD"
}


*--------------------------------------------------------------------*
Plot {
eDensity hDensity eCurrent hCurrent
Potential SpaceCharge ElectricField

eMobility hMobility eVelocity hVelocity
eQuasiFermi hQuasiFermi Potential
SRHRecombination Auger AvalancheGeneration
Doping DonorConcentration AcceptorConcentration
Current
surfaceRecombination
eIonIntegral hIonIntegral
eAlphaAvalanche hAlphaAvalanche
eTrappedCharge hTrappedCharge
HeavyIonChargeDensity
}

Physics{
Temperature=293
Mobility( DopingDep HighFieldSaturation Enormal
CarrierCarrierScattering (ConwellWeisskopf))
EffectiveIntrinsicDensity( BandGapNarrowing (OldSlotboom) )
Recombination ( SRH ( DopingDependence ) SRH(TempDependence)
eAvalanche(CarrierTempDrive) havalanche (Eparallel))
```

```
ComputeIonizationIntegrals(WriteAll)
Add2TotalDoping Tunneling (Hurkx))
}


Physics(
 MaterialInterface ="Silicon/Oxide") {
  charge (Conc=1e12)
Traps (Donor Level EnergyMid=0 fromMidBandGap conc=1.3e10
eXsection=2.9e-16 hXsection=2.9e-16 ElectricField Tunneling
(Hurkx))

* Traps (Acceptor level fromCondBand conc=5.5e14 EnergyMid=0.40
eXsection=7e-17 hXsection=7e-17 )*
Recombination(surfaceSRH)
}
```
* Surface interface traps parameters (type of trap, position of trap in
Silicon, capture cross-section of electron and holes) can be extracted
from TDRC (thermally depolarization relaxation measurements on
irradiated MOS capacitor with different X-ray doses or use different
surface recombination velocity for different X-ray doses*
* For irradiated Si detector, this charge will around 1×10^{12} to 3×10^{13}
cm^{-2} and that will depend on irradiation with X-ray doses. This effective
fixed oxide charge ($N_{ox} + N_{it}$) value can be extracted from capacitance-
voltage measurement on irradiated n-MOS test structure that is having
similar thickness of dielectric passivation layer. Or plot $1/C^2$ versus
Bias voltage for (interface trap density in cm^{-2} ($N_{it} = 0$), and $N_{ox}=0$ and
values from vendors). For the calculated value of V_{FD}, we can estimate the
($N_{ox} + N_{it}$) in the irradiated detector. *

```
Math{
  Extrapolate
  Derivatives
  NewDiscretization
  NotDamped=50
  Iterations=10
  BreakATIonIntegral
}


File {
Output = "SiD"
ACExtract="SiD"
}


{
GD trans(Pcontact1 Pcontact2 Ncontact)
Vsource_pset Vp1  (Pcontact1 0) {dc=0 }
Vsource_pset Vn   (Ncontact 0) {dc=0 sine=(0 0.1 0.01meg 0 0)}
Vsource_pset Vp2  (Pcontact2 0) {dc=0 }
}
```

* Frequency at fixed 10 KHz and may be it will vary from start to end.

```
Solve {
 Poisson
 Coupled{ Poisson Electron Hole }
# Ramp Vn and apply ac on Ncontact
Quasistationary (
     InitialStep=0.1 Increment=1.3
     MaxStep =0.1 MinStep = 9e-5
     Goal{ Paramater=Vn.dc Voltage= 0}
     ){ Coupled {Poisson Electron Hole}}
Quasistationary (
     InitialStep=0.01 Increment=1.3
     MaxStep =0.02 MinStep = 9e-5
     Goal{ Paramater=Vn.dc Voltage= 200}
)
{
ACCoupled (
StartFrequency =1e4 EndFrequency=1e4
NumberOfPoints=1 Decade
Node(Pcontact1 Ncontact Pcontact2 ) Exclude (Vp1 Vn Vp2)
)
{Poisson Electron Hole}}

 }
```

∗ **The voltage on Ncontact should be higher than full depletion voltage+50 V→ 150 V for over depleted Si detector.**

The following electrical parameters can be extracted from aforesaid TCAD program but not limited:

(a) Junction capacitance between Pcontact1 and Ncontact electrode at V_{FD} (pF/μm)
(b) Inter strip & electrode capacitances between different electrodes (see reference [6])
(c) Capacitance-voltage characteristics to get Junction capacitance at V_{FD}
(d) $1/C^2$ versus bias voltage to get V_{FD}

11.3.6 Sample File for Irradiated p+n Si Detector (Inner/ Outer) for Current-Voltage Characteristics (HEP Experiment)

11.3.6.1 TCAD CODING

- **Sample pn_des.cmd file for irradiated Si strip/pixel detector (p in n type) for current-voltage (I-V) characteristics at 20 °C (HEP experiment)**

```
Electrode{
  { name="Pcontact1"          voltage= 0.0 Areafactor=7.3e4 }
  { name="Pcontact2"          voltage=0.0}
```

```
{ name=" Ncontact "          voltage=0.0}
}
```

Area factor can be used to normalized area of strip or pixel
* File Description---*

```
File{
  Grid = "SiD_msh.tdr"
  Parameter="models.par"
  Parameter="Oxide.par"
  Plot="SiD"
  Current="SiD"
  Output="SiD"
}
```

--
```
Plot {
eDensity hDensity eCurrent hCurrent
Potential SpaceCharge ElectricField

eMobility hMobility eVelocity hVelocity
eQuasiFermi hQuasiFermi Potential
SRHRecombination Auger AvalancheGeneration
Doping DonorConcentration AcceptorConcentration
Current
surfaceRecombination
eIonIntegral hIonIntegral
eAlphaAvalanche hAlphaAvalanche
eTrappedCharge hTrappedCharge
HeavyIonChargeDensity
}

Physics{
Temperature=293
Mobility( DopingDep HighFieldSaturation Enormal
CarrierCarrierScattering (ConwellWeisskopf))
EffectiveIntrinsicDensity( BandGapNarrowing (OldSlotboom) )
Recombination ( SRH ( DopingDependence ) SRH(TempDependence)
eAvalanche(CarrierTempDrive) havalanche (Eparallel))
ComputeIonizationIntegrals(WriteAll)
Add2TotalDoping Tunneling (Hurkx) )
```

*** E5, H152K, E30K, CiOi for which fluence- done by SRH theoretical
calculations?**

**Traps ((Acceptor Level fromCondBand conc=1.24e15 EnergyMid=0.46
eXsection=3e-15 hXsection=4e-15)
(Acceptor Level fromValBand conc=3.2e12 EnergyMid=0.42
eXsection=3.05e-13 hXsection=4.1e-13)(Donor Level fromCondBand
conc=1.7e12 EnergyMid=0.1 eXsection=2.7e-15 hXsection=2e-15)(Donor
Level fromValBand conc=1.1e14 EnergyMid=0.36 eXsection=1.64e-14**

```
hXsection=2.24e-14))
}


Physics(
 MaterialInterface ="Silicon/Oxide") {
  charge (Conc=1e12)
Recombination(surfaceSRH)
}
```
* **For irradiated Si detector, this charge will around 1×10^{12} and that will depend on irradiation.** *
--
```
Math{
  Extrapolate
  Derivatives
  NewDiscretization
  NotDamped=50
  Iterations=10
  BreakATIonIntegral
}


Solve {
Poisson
Coupled{ Poisson Electron Hole }
Quasistationary (
    InitialStep=0.1 Increment=1.3
    MaxStep =0.1 MinStep = 1e-15
    Goal{ Name="Ncontact" voltage =700}
  ){ Coupled {Poisson Electron Hole} }
}
```

* **The voltage on Ncontact should be higher than calculated full depletion voltage for the given fluence.**

The following electrical parameters can be extracted from aforesaid TCAD program but not limited:

(a) Electrostatic potential
(b) E-Field as a function of the detector depth, along x-axis of the detector
(c) Space charges in the n-bulk-Si
(d) Current-voltage characteristics (current on Pcontact1/Pcontact2 as a function of Ncontact voltage), here we have to look the difference between non-irradiated and irradiated detector for generation current at V_{FD}
(e) Identification of double peak E-field or type inversion due to changes in the space charges for the different fluence
(f) Concentration of electron and hole trapped charge concentration in the bulk as a function of device depth

11.3.7   Sample File for Irradiated p+n Si Detector (Inner/ Outer) for Capacitance-Voltage Characteristics (HEP Experiment)

11.3.7.1   TCAD CODING

- **Sample pn_des.cmd file for irradiated Si strip/pixel detector (p in n type) for capacitance-voltage characteristics at 20 °C (HEP experiment)**

```
Device GD {

Electrode{
  { name="Pcontact1"          voltage= 0.0 Areafactor=7.3e4 }
  { name="Pcontact2"          voltage=0.0}
  { name=" Ncontact "         voltage=0.0}
}
```

Area factor can be used to normalized area of strip or pixel
```
* File
Description---------------------------------------------------------*

File{
  Grid = "SiD_msh.tdr"
  Parameter="models.par"
  Parameter="Oxide.par"
  Plot="SiD"
}


*---------------------------------------------------------------------*
Plot {
eDensity hDensity eCurrent hCurrent
Potential SpaceCharge ElectricField
eMobility hMobility eVelocity hVelocity
eQuasiFermi hQuasiFermi Potential
SRHRecombination Auger AvalancheGeneration
Doping DonorConcentration AcceptorConcentration
Current
surfaceRecombination
eIonIntegral hIonIntegral
eAlphaAvalanche hAlphaAvalanche
eTrappedCharge hTrappedCharge
HeavyIonChargeDensity
}

Physics{
Temperature=293
Mobility( DopingDep HighFieldSaturation Enormal
CarrierCarrierScattering (ConwellWeisskopf))
EffectiveIntrinsicDensity( BandGapNarrowing (OldSlotboom) )
Recombination ( SRH ( DopingDependence ) SRH (TempDependence)
eAvalanche(CarrierTempDrive) havalanche (Eparallel))
```

```
ComputeIonizationIntegrals(WriteAll)
Add2TotalDoping Tunneling (Hurkx))
```

E5, H152K, E30K, CiOi for which fluence- done by SRH theoretical calculations?

Traps ((Acceptor Level fromCondBand conc=1.24e15 EnergyMid=0.46 eXsection=3e-15 hXsection=4e-15) (Acceptor Level fromValBand conc=3.2e12 EnergyMid=0.42 eXsection=3.05e-13 hXsection=4.1e-13) (Donor Level fromCondBand conc=1.7e12 EnergyMid=0.1 eXsection=2.7e-15 hXsection=2e-15) (Donor Level fromValBand conc=1.1e14 EnergyMid=0.36 eXsection=1.64e-14 hXsection=2.24e-14))

```
}


Physics(
 MaterialInterface ="Silicon/Oxide") {
  charge (Conc=1e12)
Recombination(surfaceSRH)
}
```
For irradiated Si detector, this charge will around 1×10^{12} and that will depend on irradiation. *
```
*-------------------------------------------------------------------*
Math{
  Extrapolate
  Derivatives
  NewDiscretization
  NotDamped=50
  Iterations=10
  BreakATIonIntegral
}

File {
Output = "SiD"
ACExtract="SiD"
}

{
GD trans(Pcontact1 Pcontact2 Ncontact)
Vsource_pset Vp1  (Pcontact1 0) {dc=0 }
Vsource_pset Vn   (Ncontact 0) {dc=0 sine=(0 0.1 0.01meg 0 0)}
Vsource_pset Vp2  (Pcontact2 0) {dc=0 }
}
```

*Frequency at fixed 10 KHz and may be it will vary from start to end.

```
Solve {
 Poisson
 Coupled{ Poisson Electron Hole }
# Ramp Vn and apply ac on Ncontact
Quasistationary (
```

```
    InitialStep=0.1 Increment=1.3
    MaxStep =0.1 MinStep = 9e-5
    Goal{ Paramater=Vn.dc Voltage= 0}
     ){ Coupled {Poisson Electron Hole}}
Quasistationary (
    InitialStep=0.01 Increment=1.3
    MaxStep =0.02 MinStep = 9e-5
    Goal{ Paramater=Vn.dc Voltage= 200}
)
{
ACCoupled (
StartFrequency =1e4 EndFrequency=1e4
NumberOfPoints=1 Decade
Node(Pcontact1 Ncontact Pcontact2 ) Exclude (Vp1 Vn Vp2)
)
{Poisson Electron Hole}}


    }
```

* **The voltage on Ncontact should be higher than full depletion voltage+50 V→ 150 V for over depleted Si detector.**

The following electrical parameters can be extracted from aforesaid TCAD program but not limited to:

(a) Junction capacitance between Pcontact1 and Ncontact electrode at V_{FD} (pF/μm)
(b) Inter strip & electrode capacitances between different electrodes (see reference [6])
(c) Capacitance-voltage characteristics to get Junction capacitance at V_{FD}
(d) $1/C^2$ versus bias voltage to get V_{FD}

References

1. Schmitt, B., et al.: NIM-A. **501**, 267 (2003)
2. Synopsys TCAD. URL: http://www.synopsys.com/
3. Hurkx, G.A.M., Klaassen, D.B.M., Knuvers, M.P.G.: A new recombination model for device simulation including tunneling. IEEE Trans. Electron Dev. **39**(2), 331–338 (1992)
4. Poehlsen, T., Becker, J., Fretwurst, E., Klanner, R., Schwandt, J., Zhang, J.: Study of the accumulation layer and charge losses at the Si–SiO$_2$ interface in p^{C}n-silicon strip sensors. Nucl. Instr. Methods Phys. Res. A. **721**, 26–34 (2013)
5. Richter, R., Andricek, L., Gebhart, T., Hauff, D., Kemmer, J., Lutz, G., Weiss, R., Rolf, A.: Strip detector design for ATLAS and HERA-B using two-dimensional device simulation. Nucl. Instr. Methods Phys. Res. A. **377**(2), 412–421 (1996)
6. Chatterji, S., Bhardwaj, A., Ranjan, K., Namrata, Srivastava, A.K., Shivpuri, R.K.: Analysis of interstrip capacitance of Si microstrip detector using simulation approach. Solid State Electron. **47**, 1491 (2003)

Appendix

Si Detector for HEP and Photon Science Experiments

How to Design Detectors by TCAD Simulation

© Springer Nature Switzerland AG 2019

A. K. Srivastava, *Si Detectors and Characterization for HEP and Photon Science
Experiment*, https://doi.org/10.1007/978-3-030-19531-1

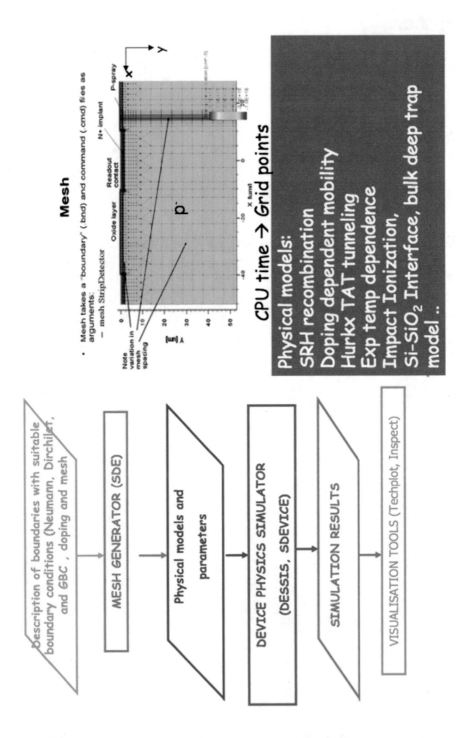

Fig. A.1.1 A Simulation flow diagram for the designing of detectors using TCAD (Technology Computer Aided Design) simulator from Synopsys

Questions on Nuclear Detectors (Background, and Modern), Accelerators and Radiation Physics (Nuclear Radiations, Hadrons)

Short Answer Type Questions

Sr. No	Questions
1	What is Alpha decay? Give one example of nuclear reaction.
2	How to detect neutron in a nuclear reaction? Give one example.
3	Write few examples of Nuclear Radiations.
4	How to shield from Alpha, Beta, Gamma, and neutron radiations?
5	Write down one radiation emitter name for Alpha, Beta, Gamma, and neutron radiations.
6	Write a few names of Photon neutron sources.
7	What is Q-Value of Nuclear reaction?
8	Write down an expression for EC.
9	Is their any energy difference between slow and fast neutron? If, yes describe.
10	What is interaction of radiation with matter? State the mechanisms.
11	What do understand by Accelerator?
12	What is the principle of G.M. counter?
13	Why Si is used in semiconductor detector?
14	How we can detect the free charge carriers?
15	Why Cherenkov detector is used?
16	What is a Linear accelerator?
17	Write down few uses of Geiger-Muller counter.
18	What is Proportional counter?
19	Write few lines about the performance characteristics accelerator.
20	What does ionization chamber?
21	What is absorbed dose?
22	Why do you need to be careful around radiation?
23	How to minimize the exposure to radiation?
24	Is radiation exposure from a nuclear power plant always fatal?

(continued)

© Springer Nature Switzerland AG 2019
A. K. Srivastava, *Si Detectors and Characterization for HEP and Photon Science Experiment*, https://doi.org/10.1007/978-3-030-19531-1

| 25 | Is nuclear radiation always harmful or are their good and helpful aspects of nuclear radiation? Explain your answer. |
| 26 | Identify the missing component.— |

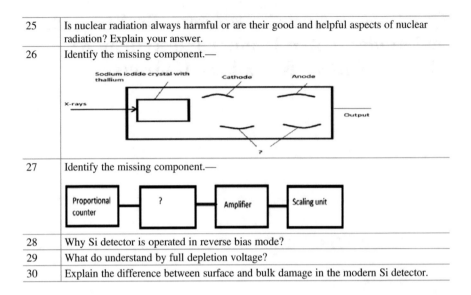

| 27 | Identify the missing component.— |

28	Why Si detector is operated in reverse bias mode?
29	What do understand by full depletion voltage?
30	Explain the difference between surface and bulk damage in the modern Si detector.

Long Answer Type Questions

Sr. No	Questions
1	Explain Beta decay nuclear reaction mechanism and its Q value, and K.E of daughter nuclei.
2	Explain stopping power.
3	How the Q-value and Kth can be extracted in any nuclear reaction? Explain with at least one example.
4	Explain Alpha decay nuclear reaction mechanism and show its Q-value.
5	Explain Gamma decay nuclear reaction mechanism using Co source.
6	Discuss about slow neutron detection.
7	Explain Beta$^+$ decay mechanism.
8	Explain biological effects of radiation using quality factor.
9	Explain any two neutron sources using nuclear reactions.
10	Discuss the radiation monitoring of Gamma radiation dose.
11	Discuss the methods used for neutron detection in detail.
12	Explain neutron detection from D-D, D-T, and photon neutron nuclear reactions in detail.
13	Explain nuclear radiations types, and the way to stopped it.
14	Discuss in detail Gamma decay mechanism through Beta decay process and energy spectrum.
15	Differentiate Q-value of Beta, Beta$^+$, EC decay.
16	Explain the heavy charge particle interaction mechanism to the matter. Show that how stopping power varies with the energy of the charged particle.

(continued)

17	Discuss in detail Alpha decay process and explain using columbic barrier tunneling.
18	Calculate the Q value and threshold energy for the nuclear reaction Ne (α, C) C? What will be the sum of K.E. of C? Given atomic masses of Ne, α, C are 19.99244, 4.00263, 12u.
19	Discuss any neutron detector.
20	Explain neutron monochromator
21	Explain the working function of semiconductor detector for the detection of neutron.
22	Discuss with one example exothermic and endothermic nuclear reaction in neutron detection.
23	Discuss the interaction of neutron radiation with matter.
24	Explain the interaction of an Alpha radiation with matter.
25	What do we mean by Braggs curve? Explain in detail.
26	Discuss stopping power Bethe Block formulae.
27	Discuss interaction of Beta radiation with matter.
28	Discuss interaction of Gamma radiation with matter.
29	Discuss neutron classification by significance of the source
30	Calculate the minimum K.E. required for nuclear reaction N(α, p)O? Given atomic masses of N, α, O and p are 14, 4, 16.9973, and 1.00769 u.
31	How can the detector detect ionizing radiation?
32	How many eh pair will be produced in Si semiconductor detector by 1 MeV radiation?
33	Why Si detector will be usually operated in 'RB mode'?
34	Discuss the types of an accelerator.
35	Explain the major differences of circular and linear collider
36	Discuss the principle of ionization chamber.
37	Explain bulk damage in the X ray semiconductor detector.
38	Discuss the signal processing schematic in general for any detector.
39	Explain surface damage in the X ray semiconductor detector.
40	Explain the detection mechanism in semiconductor detector for X-rays?
41	Write about LHC at CERN.
42	Discuss Modern CMS detector used at LHC.
43	Describe the working principle of Si strip detector.
44	Discuss the working principle of Si pixel detector.
45	Discuss X ray pixel detector design.
46	Explain X ray p+n pixel detector design.
47	Explain the Physic & Technology of X ray detector.
48	Explain the Physic & Technology of design of Si strip detector.
49	Discuss how the depletion width of the Detector increases with the bias.
50	Discuss generation current in RB p$^+$n Si detector.
51	Discuss C/V characteristics in p+n Si detector.
52	Discuss Microtron in detail.
53	Explain Electrostatic accelerators.
54	What does Synchro-cyclotron refer to? Explain.
55	What is the working principle of Cockroft—Walton generator? Elaborate it.
56	Explain Linear accelerators.
57	Discuss the Semiconductor detector design for X ray and working principle.

(continued)

58	Discuss the G.M. Counter design and working principle in detail.
59	What do we mean by Ionization chamber design? Explain with its detection method.
60	Discuss the Proportional counter schematic and its working principle.
61	The atomic masses of Eu, Sm, H, n are 151.921749, 151.919756, 1.007825, and 1.008665 u. Write a reaction, & estimate Q-value of the reaction Eu(n,p)Sm.
62	Is this reaction exothermic or endothermic? What energy does it require or give off for the following reaction Au (α, d) Hg Give that atomic mass of Au, Hg, α, d, are 196.966552, 198.968262, 4.002603, 2.014102.
63	Discuss the following (a) LHC at CERN (b) Modern detector
64	Explain the following (a) Nuclear Accelerator (b) ILC
65	Discuss the following (a) Proton synchrotron (b) Why proton synchrotron is used to produce massive particles and not e electron synchrotron?

Sample Question Bank (For Students Orientation)

This set of Multiple Choice Questions (MCQs) & Answers focuses on "Nuclear Radiation Detectors".

1. Which of the following is not a type of radiation detector?

    (a) Geiger Muller (G.M.) counter
    (b) Proportional counter
    (c) Semiconductor radiation detector
    (d) Flame emission detector
        View Answer
        Answer: d
        Explanation: Flame emission detector is not a type of radiation detector. Radiation can be detected by several methods.

2. 'When nuclear radiations pass through the counter, gas ionization is produced.' This is the principle of which of the following detectors?

    (a) Proportional counter
    (b) Flow counter
    (c) Geiger Muller counter
    (d) Scintillation counter
        View Answer
        Answer: c
        Explanation: 'When nuclear radiations pass through, gas ionization is produced.' This is the principle of which of Geiger Muller counter. It is used to measure the intensity of radioactive radiation.

3. Which of the following acts as quenching gas in G.M. counter?

    (a) Alcohol
    (b) Argon gas
    (c) Krypton
    (d) Hydrogen

© Springer Nature Switzerland AG 2019
A. K. Srivastava, *Si Detectors and Characterization for HEP and Photon Science Experiment*, https://doi.org/10.1007/978-3-030-19531-1

View Answer

Answer: a

Explanation: Alcohol acts as quenching gas in G.M. counter. It is present in a gas tight envelope along with the electrodes.

4. Which of the following acts as ionising gas in Geiger Muller (G.M.) counter?

(a) Alcohol
(b) Argon gas
(c) Krypton
(d) Hydrogen

View Answer

Answer: b

Explanation: Argon gas acts as an ionising gas in G.M. counter. It is present in a gas type envelope along with the electrodes.

5. Which of the detectors is similar to G.M. counter in construction but is filled with heavier gas?

(a) Proportional counter
(b) Strip detector
(c) Semiconductor detector
(d) Scintillation counter

View Answer

Answer: a

Explanation: Proportional counter is similar to Geiger Muller (G.M.) counter in construction but is filled with heavier gas. The output is proportional to the intensity of radiation incident on it.

6. Which of the following gases are used in proportional counter as the ionising gas?

(a) Alcohol
(b) Neon gas
(c) Krypton
(d) Heavy water

View Answer

Answer: c

Explanation: Proportional counter is filled with krypton. It acts as an ionising gas.

7. Which of the following is the main disadvantage of solid state semiconductor radiation detector?

(a) Low accuracy
(b) Low sensitivity
(c) It should be maintained at low temperature
(d) High avalanche breakdown voltage

View Answer

Answer: c

Explanation: The main disadvantage of solid state semiconductor radiation detector is that it must be maintained at low temperature. This is necessary to reduce noise and to prevent deterioration of detector characteristics.

8. Scintillation detector is a large flat crystal of which of the following materials?

   (a) Sodium chloride
   (b) Sodium iodide
   (c) Sodium sulphate
   (d) Sodium carbonate

     View Answer

     Answer: c

     Explanation: Scintillation detector is a large flat crystal of sodium iodide. It is coated with thallium doping.

9. When X-ray enters the solid state (Si) detector it produces ion pair rather than electron-hole (e-h) pair.

   (a) True
   (b) False

     View Answer

     Answer: b

     Explanation: When X-ray enters the solid state detector it produces electron-hole pair rather than ion pair. The output signal is taken from an ohmic contact (Al layer).

10. Which of the following materials are used as the insulation between inner and outer electrodes of the ion chamber?

   (a) Ceramic
   (b) Plastic
   (c) Polytetrafluoroethylene
   (d) Polyacrylamide

     View Answer

     Answer: c

     Explanation: Polytetrafluoroethylene is used as the insulation between inner and outer electrodes of the ion chamber. The material has very high resistance.

11. Liquid samples must be counted using ionization chamber by placing them in which of the following?

   (a) Test tube
   (b) Curvette
   (c) Ampoules
   (d) Faraday cup

View Answer

Answer: c

Explanation: Liquid samples must be counted using ionization chamber by placing them in ampoules. The ampoules are placed in the chamber.

12. Gaseous compounds containing radioactive sources can be directly introduced into the ionization chamber.

   (a) True
   (b) False

   View Answer

   Answer: a

   Explanation: Gaseous compounds containing radioactive sources can be directly introduced into the ionization chamber. Liquid samples cannot be introduced directly.

13. Liquid Scintillators are used for which of the following materials?

   (a) Low energy beta materials
   (b) High energy beta materials
   (c) Low energy gamma materials
   (d) Fast neutron

   View Answer

   Answer: a

   Explanation: Liquid Scintillators are used for low energy beta materials. Solid scintillators are used for high energy beta materials.

14. Given below is the block diagram of proportional counter. Identify the unmarked component.

   (a) Collimator
   (b) Detector crystal
   (c) Pre-amplifier
   (d) Shaper

   Answer: c

   Explanation: The unmarked component is pre-amplifier. There are two amplifiers namely pre-amplifier and main amplifier.

15. Given below is a diagram of Scintillation detector. Identify the unmarked component.

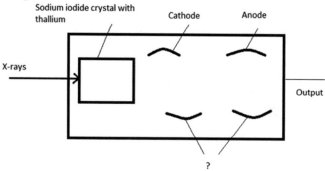

Sodium iodide crystal with thallium

Cathode

Anode

X-rays

Output

?

(a) Lens
(b) Collimator
(c) Dynodes
(d) X-ray tube

Answer: c

Explanation: The unmarked components are dynodes. Scintillation detector is a combination of scintillator and photo multiplier tube.

Printed in the United States
By Bookmasters